# Fact and Feeling

SCIENCE AND LITERATURE
A series edited by George Levine

# Fact and Feeling

*Baconian Science and the Nineteenth-Century Literary Imagination*

JONATHAN SMITH

The University of Wisconsin Press

The University of Wisconsin Press
114 North Murray Street
Madison, Wisconsin 53715

3 Henrietta Street
London WC2E 8LU, England



5 4 3 2 1

Printed in the United States of America

Library of Congress Cataloging-in-Publication Data
Smith, Jonathan, 1960–
Fact and feeling: Baconian science and the nineteenth-century literary imagination / Jonathan Smith.
288 p. cm. — Science and literature)
Includes bibliographical references and index.
ISBN 0-299-14350-3 ISBN 0-299-14354-6 (pbk.)
1. English literature—19th century—History and criticism.
2. Literature and science—Great Britain—History—19th century.
3. Bacon, Francis, 1561–1626—Influence. 4. Science in literature.
I. Title. II. Series.
PR468.S34S65 1994
820.9′008—dc20 94-14014

For my parents, with love and thanks

# Contents

Acknowledgments ix

Introduction 3

1. The Science of Science: Baconian Induction in Nineteenth-Century Britain 11
2. Reweaving the Rainbow: Romantic Methodologies of Poetry and Science 45
3. Seeing through Lyell's Eyes: The Uniformitarian Imagination and *The Voyage of the Beagle* 92
4. The "Wonderful Geological Story": Uniformitarianism and *The Mill on the Floss* 121
5. Ruskin's "Analysis of Natural and Pictorial Forms" 152
6. "Euclid Honourably Shelved": Edwin Abbott's *Flatland* and the Methods of Non-Euclidean Geometry 180
7. "The Methods Have Been of Interest": Sherlock Holmes, Scientific Detective 211

Notes 241
Index 271

# Acknowledgments

This book began as a dissertation for the Department of English and Comparative Literature at Columbia University. My chief thanks are of course due to my sponsors, Jonathan Arac and John Rosenberg, each of whom played the role of mentor with thoroughness and grace. For their insightful comments and questions, their advice and support, I am grateful.

Other readers to whom I am indebted for assistance and many kindnesses include Robert Patten, Carolyn Williams, Susan Winnett, Karl Kroeber, Sydney Morgenbesser, and David Albert. Lee Sterrenburg provided a thorough and helpful review of the manuscript for the University of Wisconsin Press. To graduate school colleagues Beth Harrison, Suvendi Perera, Steve Greenfeld, Lisa Gitelman, Liz Auran, and Garrett White I owe not only critical readings and thoughtful discussion, but also encouragement and support. To George Levine in his various roles as mentor, editor, reader, and scholar, I offer special thanks.

I am pleased to acknowledge the support of the National Endowment for the Humanities in the form of a 1992 Summer Seminar, "Methodological Debates in Nineteenth-Century Physics," directed by Peter Achinstein at the Johns Hopkins University. Research on Chapter 5 was begun at that seminar, and I am grateful to Professor Achinstein, his graduate assistant, Laura Snyder, and my fellow seminar participants for their feedback. I would also like to thank the Elliott Dobbie Fund and the Department of English and Comparative Literature at Columbia for a grant supporting the preparation of the manuscript.

An earlier version of Chapter 4 originally appeared in *Papers on Language and Literature* 27 (1991): 430-52, and an earlier version of Chapter 5 is forthcoming in a volume of essays edited by Peter Achinstein for Krieger Publishing Company.

Finally, I thank Jackie Lawson for love and friendship that have healed wounds: "I were but little happy if I could say how much."

# Fact and Feeling

# Introduction

Shortly after the publication of Darwin's *Descent of Man* in 1871, there appeared in *Blackwood's* a humorous poem about Darwin's views on the origin of the human species:

THE DESCENT OF MAN.
A CONTINUATION OF AN OLD SONG.
Air—"Greensleeves."
(Darwin loquitur.)

"Man comes from a Mammal that lived up a tree,
And a great coat of hair on his outside had he,
Very much like the Dreadnoughts we frequently see—
Which nobody can deny."[1]

When the voice of Darwin completes its eight-stanza explanation of man's descent, the poem's narrator turns his attention to a critique of that explanation. The criticisms offered are the standard ones: there's nothing in Darwin's theory that hasn't been anticipated by either Lucretius or Erasmus Darwin; the tracing of man's genealogy merely pushes back the question of origins rather than supplying an explanation; natural selection cannot account for the presence of man's soul or his moral sense. But the poem's parting criticism, also a standard one, aims not at the content of Darwin's theory but at the method used to obtain it:

I would ne'er take offence at what's honestly meant,
Or that truth should be told of our lowly descent;
To be sprung from the dust I am humbly content—
Which nobody can deny.
But this groping and guessing may all be mistaken,
And in sensitive minds may much trouble awaken,
So I'll shut up my book, and go back to my *Bacon*—
Which nobody can deny.

Darwin's theory of descent amounts to "groping and guessing." It violates the most basic principles of Baconian induction, and therefore cannot really be considered scientific at all. As a theory it is ingenious, but it is mere speculation, the product of an individual's fertile imagination rather than the result of a gradually ascending series of inductions grounded in observed facts. Lest there be any doubt about this, the poet includes a footnote after his reference to Bacon: "Certainly the Darwinian theory, though it may be interesting as a theory, is a considerable encroachment on Baconian principles, which require that no theory should be adopted without an adequate induction from facts much more directed and complete than any that the Darwinians have yet discovered—if, indeed, they have discovered any fact at all that infers the possibility of the transformations which they promulgate."

This confident invocation of Bacon and Baconian induction came at a time of skepticism about the value of "Baconian principles," for by 1871 the long and ongoing debate over scientific method in the nineteenth century had led to the revision if not to the outright repudiation of the methodological specifics of the *Novum Organum*. Fourteen years earlier, Robert Leslie Ellis had declared, in what remains the standard edition of Bacon's *Works*, that despite Bacon's importance to the spirit of scientific investigation, the details of his inductive method were "practically useless." Pure objectivity, absolute certainty, avoidance of hypotheses, gradually widening generalization, systematic elimination of possible explanations—all these features of Baconian induction were declared to be inadequate or even impossible representations of how a scientist works. Science, as Ellis and countless others argued, requires hypothesis and deduction, the genius of an individual creative mind willing to guess, to imagine, to leap to conclusions.

Yet the poem's attack on Darwin's method signals both the continued authority of Bacon and the scope of the debate about method itself. In the arguments that swirled around the century's most controversial scientific theory, the issue of Darwin's method assumed, as it did in the *Blackwood's* poem, a place of prominence. G. H. Lewes wrote a series of articles for the *Fortnightly* entitled "Mr. Darwin's Hypotheses." John Stuart Mill assigned Darwin's method a place in his *Logic*. T. H. Huxley defended Darwin from those like the *Blackwood's* poet who "prate learnedly about Mr. Darwin's method, which is not inductive enough, not Baconian enough, forsooth, for them." Yet it was an issue not merely for scientists and philosophers, but for a "highbrow" magazine's parody of a "lowbrow" cultural form. It was an issue broached as readily in the witty stanzas of the poem as in the essayist's prose of the footnote. It was an issue of cultural importance that surfaced relentlessly in the productions of novelists, scientists, logicians, philosophers, historians, and poets.

The nineteenth-century debate over scientific method included much more than a discussion of the logical process of scientific discovery. It could not avoid basic epistemological questions about reality and the relationship between observer and observed; it evaluated the nature and value of scientific knowledge vis-à-vis other forms of knowledge, including supposedly more intuitive forms like religion and poetry; it contributed to the rewriting of the history of science and of philosophy; it touched on the place of science within society, its role in education, in industry, and in social policy. The debate over scientific method, though not the only locus in the relationship between science and the culture at large, is certainly an important locus through which this matrix of connections may be traced.

This study begins, therefore, with the assumption that science is a form of cultural discourse: like literature or history or music or art or religion, it both shapes and is shaped by the culture of which it is a part. This is not to deny that science has in our culture obtained a special status, nor is it an effort to demean science by claiming that it is "merely" literary, a position that demeans literature as much as science. Rather, it is the recognition that this elevation of scientific discourse—which largely took place during the nineteenth century—does not also isolate it from, or make it necessarily antithetical to, other forms of discourse. "Science" and "literature" are not cultural monoliths but diverse and always changing, like the culture itself.

Studies of science and literature over the last decade or two have tacitly assumed or triumphantly asserted that the two cultures debate is over, that there is one culture, not two. Rather than lamenting that scientists and humanists don't and can't talk to each other, we now demonstrate that it hasn't always been so, that it needn't be so today, and that it usually isn't so today even when we think it is. Such a view is undoubtedly right—or at least more productive.[2] Yet within this new approach may often be found quietly lurking the same assumption that animated the earlier lamentations. So deeply rooted is our sense of some fundamental difference between science and literature that even as we demonstrate the speciousness of such a notion, we continue, implicitly, to employ it. We know that the Romantics were not anti-scientific, and we know that the opposition of the subjective, organic, imaginative artist and the objective, mechanical, unimaginative scientist is inadequate. Yet when we find the literary and the scientific interpenetrating, we often seem surprised. We talk about the juxtaposition of opposites, and our discussion assumes that even such juxtaposition is unusual.

Sherlock Holmes is an excellent example. He is commonly seen as a remarkable character because he combines in one person what are thought to be two different personalities, romantic artist and impersonal scientist. As we celebrate the presence of these personalities in one person, however, we do so in a way that ultimately reinforces the schizophrenia we are denying. Rather than recognizing the presence of the scientific in the "artistic temperament" and the artistic in the "scientific temperament," we describe Holmes not as a single, integrated personality, but as two very different personalities (Holmes the scientist on a case, Holmes the artist in a cocaine-induced reverie) that cannot be expressed simultaneously. Watson falls into this view initially, but he grows out of it, and so must we. Artist and scientist coexist in Holmes's character because art and science partake of each other's aims, language, and methods.

What is required instead is the study of specific points of contact, of actual scientists and writers, actual works of science and literature, rather than "science" and "literature." If this book has a manifesto, it may be found in the words of George Levine, who declares that "[p]articipating equally but differently in the culture's myths and ideologies, science and literature support, reveal, and test each other."[3] With the caveat that the participation of science and literature in any culture's myths and ideologies is not always equal, my endeavor here is to explore some specific instances of this relationship in the science and literature of nineteenth-century Britain, with the hope that what results is a more productive way of describing our own "one culture."

The "ands" of my title are thus meant to suggest the fluidity rather than the dichotomy of this relationship. I am borrowing from the critic Edward Dowden's 1877 essay "The Scientific Movement and Literature":

> Any inquiry at the present day into the relations of modern scientific thought with literature must in great part be guided by hints, signs, and presages. The time has not yet come when it may be possible to perceive in complete outline the significance of science for the imagination and emotions of men, but that the significance is large and deep we cannot doubt. Literature proper, indeed, the literature of *power,* as De Quincey named it, in distinction from the literature of knowledge, may, from one point of view, be described as essentially non-

scientific, or even antiscientific. To ascertain and communicate facts is the object of science; to quicken our life into a higher consciousness through the feelings is the function of art. But the knowing and feeling are not identical, and a fact expressed in terms of feeling affects us as other than the same fact expressed in terms of knowing, yet our emotions rest on and are controlled by our knowledge.[4]

The terms of this Victorian rumination are set by the Romantics. Dowden's assertion that "the time has not yet come" to assess the significance of science for the imagination seems a conscious answer to Wordsworth's comment in the 1802 preface to the *Lyrical Ballads* that "[i]f the time should ever come when what is now called science, thus familiarized to men, shall be ready to put on, as it were, a form of flesh and blood, the poet will lend his divine spirit to aid the transfiguration, and will welcome the being thus produced, as a dear and genuine inmate of the household of men." Yet Dowden implies that if a complete assessment is not within our grasp, we already know enough to be certain that the significance is large and deep. In a similar fashion, Dowden both appropriates and revises the De Quinceyan distinction between the literature of knowledge and the literature of power, a distinction De Quincey claimed to base on years of conversation with Wordsworth. Dowden rightly notes that De Quincey's distinction tends to create a schism between science and art; the literature of power is accorded a superior status that can be seen as non- or even antiscientific. He goes on, however, to quietly challenge De Quincey's assumptions by restoring some priority to the literature of knowledge. For Dowden, art also depends on fact. It expresses fact differently—its methods and goals are affective as opposed to science's discursive drive to "ascertain and communicate"—but its imaginative, emotional base is itself grounded on and shaped by knowledge. And yet the function of art is to quicken our life into a *higher* consciousness than science's expression of fact in terms of knowing can afford.

Dowden's shifts are in many ways paradigmatic of the Victorian sense of the relationship between science and literature. Crucially, the roots of that shifting sense lie in Romantic views that are themselves fractured and unstable and capable of varied interpretation and appropriation. The effort to distinguish differences in the two enterprises, in their means and their ends, while simultaneously maintaining a connection between them as parts of a unified scheme of human knowledge, is a difficult balancing act. Dowden at various moments is thus drawn to very different possibilities: that science is completely disconnected from feeling but that its expression of fact as knowledge is ultimately the basis for art's expression of fact as feeling; that art's feeling is both higher than and dependent on science's knowing; that science and art can be rigorously separated but that science's influence on art is vast and deep. Indeed, we see in Dowden the strands of thought that could and often did lead to monolithic articulations of the relationship, but we would be wrong to take any monolithic conception, either that the Victorians saw science and literature as compatible or that they saw them as fundamentally at odds, as the dominant cultural view. The dominant cultural view is the lack of a dominant cultural view, the tensions and controversies rather than the univocal expressions of certainty that have perhaps too often been attended to.

Chapter 1 provides a historical and thematic frame for the body of the book by tracing in detail the nineteenth-century debate over scientific method. I concentrate on the assessments of Baconian induction and the Baconian-Newtonian tradition, devoting special attention first to the rhetoric used to define the "naive Baconianism" that was the target for attack, and then to the ambivalence inherent in trying to retain some authority for Bacon's contribution to scientific method while altering or rejecting the details of that contribution. And I demonstrate that the growing emphasis on the imagination and creativity of the scientist—what Tyndall called "the scientific use of the imagination"—brings scientific method into close accord with the methods of the literary imagination.

Chapters 2 through 7 pair a particular scientist and/or scientific theory with a particular writer and/or a particular work. Chapter 2 sets the stage for Victorian tensions by exploring the Romantic tensions in the poetics of Wordsworth, the more general comments on method in Coleridge's *The Friend*, and the scientific methodology of Humphry Davy. The prefaces to the *Lyrical Ballads* and the 1815 *Poems* indicate that Wordsworth is rejecting naive Baconianism in both poetry and science, and that Wordsworth's sense of how a poet produces poetry is very similar to Davy's sense of how a scientist goes about doing science, although each wants ultimately to privilege his own endeavor. Coleridge goes beyond both men by actively and consciously redefining the Baconian tradition in his search for a method that will apply to all forms of knowledge, by embracing tensions and contradictions that made Wordsworth and Davy uneasy. In the process I consider how the language of these early Romantics came to be appropriated in subsequent arguments for the disjunction of literature and science, and I argue that much recent criticism continues to enforce this distinction even as it tries to explode it.

In Chapters 3 and 4 I take up geology, arguably the century's most popular and accessible science, and narratives of personal history. My focus is on uniformitarianism, the influential geological theory articulated by Darwin's mentor, Charles Lyell. In Chapter 3, I discuss Darwin's account of his journey on the *Beagle*, first published just six years after the completion of Lyell's *Principles of Geology* (1830–33), as a narrative in which not only geological observations but literary devices are shaped by Lyell's rhetoric. Lyell presented uniformitarianism as simultaneously inductive and imaginative—indeed, as more imaginative than its catastrophist rivals. Darwin thus uses his carefully cultivated uniformitarian imagination to convert the "romance" of catastrophist geology into the "realism" of uniformitarianism, which, in its emphasis on the extraordinariness of the ordinary, also has much in common with Wordsworthian Romanticism. In Chapter 4, this tradition is extended to George Eliot's *The Mill on the Floss* (1860), written when Lyell's uniformitarianism was at its height and Darwin's theory of natural selection was newly published. I argue that Eliot's novel is profoundly uniformitarian, that it offers a Lyellian reading of Darwin and of the concluding flood, deliberately subverting the catastrophist geology and natural theology of men like Richard Owen and William Buckland. Yet the dissatisfaction experienced by readers at the novel's end is not unreasonable, for it exposes some of the theoretical and methodological difficulties contained both within uniformitarianism itself and within Eliot's realist project.

Chapters 5 and 6 examine responses from mid- and late-century to more aggressive empiricist championing of the scientific imagination. In my discussion of John Ruskin's criticism of art and science in Chapter 5, I present Ruskin's methodology for the depiction of nature in *Modern Painters* (1843–60) as basically consistent with, although somewhat more conservative in its emphasis than, that of his scientific contemporaries. His later works, with their more strident criticism of materialist science, are still based on that same methodology, although developed to provide an alternative to the threat posed by Tyndall's marriage of materialism and speculative imagination. In Chapter 6, I position Edwin Abbott's *Flatland* within the debate over the methodological validity and philosophical implications of non-Euclidean geometry. Although Abbott's book has generally been treated as very liberal and forward-looking on a variety of fronts, I contend that Abbott is unwilling to embrace the radical implications endorsed by the new geometry's most aggressive empiricist popularizers, and that this unwillingness is closely related to his own methodological views in matters of science and religion.

My discussion of Sherlock Holmes in Chapter 7 sees in the methods of the great detective a retrospective encapsulation of the century-long debate, a testimonial to the power of an inductive methodology charged with the speculative and often mysterious processes of hypothesis, intuition, and deduction. Returning to the issues raised with the Romantics, I contend that what we have in Holmes is not the surprising juxtaposition of disparate traditions but the tensions evolved in the incorporation of the scientific into the artistic and the artistic into the scientific. We can see this by comparing the aesthetics of Oscar Wilde with the works of the scientists most influential on Conan Doyle: Huxley and Tyndall.

My approach attempts to provide a sense of the diversity with which science engaged nineteenth-century Britain in general, literary artists in particular. I have deliberately chosen to study over the course of the century a number of different scientists in a variety of scientific disciplines as well as a number of different writers in several different genres. This range is designed to counter the notion unconsciously fostered by studies of one author (for example, Eliot) or one scientist (for example, Darwin) that a writer's interest in science or a scientist's influence outside the scientific community is unusual, limited to the extraordinary mind. By showing that writers and scientists both participated in the debate over Baconian induction, that they drew on the language and methods of each other's discourses, I hope to offer a glimpse (to steal metaphors from both Eliot and Darwin) of that web of affinities, that entangled bank, which is the relationship of science and literature in nineteenth-century British culture.

# 1

# The Science of Science

## Baconian Induction in Nineteenth-Century Britain

If there is any science which I am capable of promoting, I think it is the science of science itself, the science of investigation—of method.

—John Stuart Mill, 1831

The art which Bacon taught was the art of inventing arts.

—Thomas Macaulay, 1837

The greatest invention of the nineteenth century was the invention of the method of invention.

—Alfred North Whitehead, 1925

In Thomas Sprat's 1667 *History* of the recently founded Royal Society appears Cowley's famous ode "To the Royal Society." Cowley eulogizes Francis Bacon as the man responsible for liberating "Philosophy" (science) from blind obedience to the views of the ancients, and for providing it with a method (induction) by which to reform itself and to provide access to a Promised Land of hitherto undiscovered scientific truths:

> Bacon, at last, a mighty man, arose,
> Whom a wise King and Nature chose
> Lord Chancellour of both their laws,
> And boldly undertook the injur'd Pupils [Philosophy's] caus. . . .
> . . . . . . . . . . . . . . . . . . . . . . . . . . . . . . . . . . . . . . . . . . .
> Bacon, like Moses, led us forth at last,
> The barren Wilderness he past,
> Did on the very Border stand
> Of the blest promis'd Land,
> And from the Mountains Top of his exalted wit,
> Saw it himself, and shew'd us it.[1]

During the eighteenth century, the name of Isaac Newton became coupled with that of Bacon in the genealogy of the chosen people of British science. Bacon was Moses, the lawgiver in the *Novum Organum* of the new inductive philosophy, but a man who failed to put that philosophy successfully into practice, a man who brought his people to the edge of the Promised Land but did not himself enter in. Newton, however, fulfilled the promise of the Baconian laws and took possession of the world of nature in its widest sense, from the terrestrial to the celestial. By 1808, therefore, Humphry Davy could speak in a public lecture about the "legitimate practice" of science as "that sanctioned by the precepts of Bacon and the examples of Newton." In his 1820 inaugural address as president of the Royal Society, Davy urged his colleagues that their work "be guided by the spirit of philosophy, awakened by our great masters, Bacon and Newton, that sober and cautious method of inductive reasoning, which is the germ of truth and of permanency in all the sciences."[2]

A similarly reverential assessment of Bacon and his methodology appeared in the century's first edition of the *Encyclopaedia Britannica*. In his "Dissertation: Exhibiting a General View of the Progress of Mathematical and Physical Science, Since the Revival of Letters in Europe," John Playfair declared that as Bacon "has had no rival in the times which are past, so is he likely to have none in those which are to come. . . . If a second Bacon is ever to arise, he must be ignorant of the first."[3] Yet by the ninth edition of the *Encyclopaedia* in 1878, Bacon's method was unceremoniously dismissed:

> It has been pointed out, and with perfect justice, that science in its progress has not followed the Baconian method; that no one discovery can be pointed to which can be definitely ascribed to the use of his rules, and that men the most celebrated for their scientific acquirements, while paying homage to the name of Bacon, practically set at nought his most cherished precepts. . . . The inductive formation of axioms by a gradually ascending scale is a route which no science has ever followed, and by which no science could ever make progress. The true scientific procedure is by hypothesis followed up and tested by verification; the most powerful instrument is the deductive method, which Bacon can hardly be said to have recognized.[4]

A generation later, the demise of Baconian induction had become a matter of historical curiosity for professional philosophers: "why," asked one of them, "is Bacon's method of science dead?"[5]

As the differences in these views suggest, definitions and assessments of Baconian induction—and of scientific methodology in general—underwent significant changes during the course of the nineteenth century. Henry Hallam, writing in the 1830s, argued that the copious reference to Bacon in both popular and more specialized works of science and philosophy "is not much older than the close of the last century. . . . I should expect that more have read Lord Bacon within these last thirty years than in the preceding two centuries."[6] As something of a Bacon supporter, Hallam was perhaps reluctant to note that many of those reading and referring to Bacon were doing so in a spirit of criticism and revision, but no such reluctance is evident in Stanley Jevons' 1874 assessment that "[w]ithin the last century a reaction has been setting in against the purely empirical procedure of Francis Bacon."[7] As the century progressed, contemporaries were aware that the unprecedented interest in Bacon's methodology had taken a negative turn. Reverence for Bacon and for the general tenor of his thought continued, but the details of his inductive philosophy were criticized with increasing boldness. Thomas Fowler's 1878 edition of the *Novum Organum* included both an extensive bibliography of the various commentaries on Bacon's methodological treatise and an admission that he was attributing more value to Bacon's methods "than is usually conceded to them."[8] In 1905, John Robertson provided a similar summary of the "debate which has circled round the name of Bacon . . . in the past century" but could find none of the methodological value that Fowler had argued for: according to Robertson, the "real triumph" of Bacon lay not in his precepts but in his prose style.[9]

## "THE COMMON NOTION OF LORD BACON"

What happened to scientific methodology in nineteenth-century Britain? Speaking in broad terms, the shift from Baconian induction to something like what we now call the hypothetico-deductive method. Why is this shift important for a study of nineteenth-century British literature? Because this new formulation of scientific method consciously sought to portray science as an imaginative, speculative, creative enterprise. Science did not abandon its claims to be an objective and authoritative pathway to truth, but it did assert more openly that this truth is obtained through, rather than at the expense of, the creative imagination. "Baconianism" was seen as too sterile, too mechanical, and too impersonal to capture the artistic quality of the scientist doing science.

This chapter endeavors to fill in the outline that has already been sketched, to bridge the two endpoints. In the process, I hope to show that the debate over scientific method was not confined to a narrow community of scientists or philosophers, but engaged virtually the entire culture. Many of the people involved in this discussion—Coleridge, Macaulay, J. S. Mill, William Whewell, G. H. Lewes, Stanley Jevons, and T. H. Huxley, to name but a few—were obviously involved in a variety of different disciplines and hence able to spread the issues well beyond whatever parochial origins might have initially obtained in a given field.

But what *is* Baconian induction? We needn't go to Bacon to answer this question. Although I shall refer to Bacon's works from time to time, my focus will be not on what Bacon said, but on what nineteenth-century writers said he said. We don't need to read Bacon (or Newton) to talk about what the nineteenth century thought of him. Indeed, reading Bacon directly is in one sense counterproductive, for it can distract us from the particular elements that different nineteenth-century commentators highlight and emphasize. The nature and value of Bacon's methodology, already interpreted and appropriated in particular ways in the seventeenth and eighteenth centuries, were hotly contested in nineteenth-century British culture. For example, while Lord Brougham applauded Bacon and Newton for not giving way to hypotheses, William Whewell said hypotheses were crucial and pointed to the places where Bacon and Newton employed them, and Stanley Jevons claimed that Newton's awareness of the value of hypotheses made Newton rather than Bacon the true founder of scientific method. Similarly, Bacon's own application of his inductive method to the question of heat had been considered a rather embarrassing failure for years, but when the development of thermodynamics established that heat was, as Bacon had suggested, a form of motion, the example was employed as further evidence of Bacon's prophetic insight.

In *Science in Culture*, Susan Cannon argues that "Baconianism" is the creation of modern historians of science. While it is easy to find nineteenth-century figures bemoaning the presence of a Baconian orthodoxy, she claims that it is virtually impossible to find anyone who supports this supposed orthodoxy.[10] But such an account seems inadequate, for it is not difficult to find endorsements of Baconian induction. In his *Elements of Chemical Philosophy* (1812), Davy credits Bacon with the development of "the general system for improving natural knowledge": "Till his time there had been no distinct views concerning the art of experiment and observation. . . . He taught that Man was but the servant and interpreter of Nature; capable of discovering truth in no other way but by observing and imitating her operations; that facts were to be collected and not speculations formed; and that the materials for the foundations of true systems of knowledge were to be discovered, not in the books of the ancients, not in metaphysical theories, not in the fancies of men, but in the visible and tangible external world."[11] Davy is here attempting to provide the relatively young science of chemistry with the status that comes from locating it squarely within the Baconian tradition. The method of chemistry is the method of Bacon: observation and experiment, the collection of facts rather than the forming of speculations. Chemistry is one of the "true systems of knowledge" of a natural world that is "visible and tangible," not a tissue of metaphysical theories founded only on "the fancies of men." And Bacon really is the first man to provide this method, to liberate philosophers from metaphysics by propounding the "art" of observation and experiment.[12]

A more complex rendering of the historical situation sees the status of Bacon and Baconian induction as a locus for genuine debate and controversy. Richard Yeo argues that Bacon's ideas "sponsored various interpretations and, in turn, different intellectual and political uses"; Antonio Pérez-Ramos echoes this in his discussion of "the meanings of Baconianism."[13] For Pérez-Ramos, British science in the first half of the nineteenth century "had to be ceremonially Baconian if it aspired to respectability."[14] This ceremonial quality had its roots in the historical tradition outlined above, an emphatically British tradition that had enshrined Bacon as its methodological master and Newton as his successful follower, and that had melded the study of nature with the worship of God. Yet, at the same time, there was more at stake than ceremony. As Pérez-Ramos elsewhere shows and as Yeo stresses, the expositions of the inadequacies of a "common" or "vulgar" or "naive" view of Bacon had as their goal the dissociation of Bacon's methodology from French Enlightenment thought, with its perceived celebration of materialism, atheism, and radical politics.[15] This was, of course, not simply an issue of national boundaries or traditions. Whewell, for example, as a member of the Oxbridge Anglican elite, had different social, political, religious, and philosophical axes to grind than did the utilitarian, reform-minded Mill. The debate over Baconianism, in other words, touched on many of the culture's most deeply held beliefs and most explosive areas of controversy.

It is not the purpose of this chapter to chart these competing interests. Rather, I want to focus on areas of agreement in the efforts to construct, in roughly the first two-thirds of the century, the "proper" Baconianism that was to replace its "naive" counterpart, efforts that led in the later years of the century to a discrediting of Baconianism in almost any specific methodological form. Both Yeo and Pérez-Ramos suggest that the decline in Bacon's reputation as a methodologist had its roots in part in the work of those nineteenth-century thinkers who admired him, and this chapter will support that claim through an analysis of the language of numerous participants in the Baconian debate.[16] And, most important for my concerns, both the competing interests and the overall areas of agreement help to explain why scientific methodology also interested nineteenth-century literary figures. With controversial intellectual and political concerns connected to the narrow question of Bacon's methodology, literary figures could not help but take an interest. Even more directly, not only had Bacon himself been seen as a literary figure, but one of the central areas of agreement in the methodological debate, as I suggested above, had to do with the need for and the place of the imagination in science. Nineteenth-century literary figures were happy with the denigration of a methodology that they tended to see as hostile to their own, even as they presented aspects of their methodology as "properly" Baconian.

Objections to "naive" Baconianism are directed particularly against the notion that the indiscriminate collection of facts will lead eventually and inevitably to the formation of a theory. In an 1819 lecture, for example, Coleridge says the "common notion of Lord Bacon" is "that you are to watch everything without having any reason for so doing, and that after you have collected the facts that belong to any subject . . . you may proceed to the theory." He criticizes this "common notion" on a variety of points: that you must have a preliminary theory before you can observe, that it is impossible to collect all the facts, and that theories do not magically arise out of facts but are the result of the imaginative interpretation of facts. The "true Baconian philosophy," says Coleridge, consists in "a profound meditation on those laws which the pure reason in man reveals to him, with the confident anticipation and faith that to this will be found to correspond certain laws in nature."[17]

Similar complaints are voiced by Whewell in his *Philosophy of the Inductive Sciences* (1840), the century's most unashamedly idealist philosophy of science in Britain. Whewell rejected empiricist claims that scientific knowledge was ultimately based on sense impressions, arguing instead that Fundamental Ideas like causation, polarity, and substance are necessary truths essential to making sense of empirical "facts." Thus Whewell contends that the common conception of Bacon as having raised observation and experiment at the expense of ideas—

the very notion offered by Davy—is dangerously misleading. Such a view tells only half the truth, says Whewell, for Bacon gave "due attention to Ideas, no less than Facts"—or at least he would have, had he completed his system. This mania for facts has also led the popular mind to overemphasize the utility of scientific knowledge (Bacon's Experiments of Fruit) when "an acquaintance with the laws of nature is worth having for its own sake" (Bacon's Experiments of Light).[18] For Whewell, developing an updated version of Bacon's method required the demolition of the unsophisticated and overly utilitarian common view that had acquired such sway.

Attacks on the Baconian emphasis on facts were not limited to those thinkers like Coleridge and Whewell influenced by German idealism. In his exposition of Comte's positive philosophy, G. H. Lewes ridiculed those Baconians who prated about the importance of facts and who incorporated Newton into the Baconian program: "Newton's assertion, *Hypotheses non fingo*—make no hypotheses—has been incessantly repeated by men who fancy themselves Baconian thinkers when they restrict their incompetence to what they call 'facts.' No reader of these pages need be told that such ideas of science are utterly irrational."[19] Though positivism was too empirical even for such empirically minded people as Mill and Huxley, it, too, sought to distance itself from what it saw as the distorted popular interpretation of the virtues of Baconian fact collecting.

As for Davy's claim, also inscribed in Cowley's ode, that Bacon had liberated science from metaphysics, providing a new method of experiment and observation, Macaulay called it "[t]he vulgar notion about Bacon." The idea that Bacon "invented a new method of arriving at truth, which method is called Induction; and that he exposed the fallacy of the syllogistic reasoning which had been in vogue before his time" was "extravagant nonsense."[20] In Macaulay's view, Bacon did not invent induction, nor was he the first to challenge the supremacy of the syllogism. Furthermore, induction is just a fancy name for common sense, something every person practices every day. Although Macaulay's essay was celebrated for its elevation of the empirical over the idealist tradition, it is an excellent example of the tensions that occur in nineteenth-century assessments of Bacon. The bulk of the essay is critical of Bacon the man; much of its energy is directed toward correcting the too-favorable account of Bacon's character in Basil Montagu's recent edition of Bacon's *Works*. Yet, in spite of his criticisms of man and method, Macaulay also offers considerable praise of the *Novum*

*Organum*. He declares that "[n]o book ever made so great a revolution in the mode of thinking—overthrew so many prejudices—introduced so many new opinions."[21]

Macaulay's claims occasioned considerable debate, sending friend and foe alike not just to the writings of Bacon, but to those in the intervening two centuries who had written about the writings of Bacon. Like Hallam, they found that the eighteenth century in particular had done much to create the Baconianism that now merited rejection or revision.[22] Jevons claimed that eighteenth-century science was deluding itself if it thought that its methods were Bacon's: "Throughout the eighteenth century science was supposed to be advancing by the pursuance of the Baconian method, but in reality hypothetical investigation was the main instrument of progress. It is only in the present century that physicists began to recognize this truth."[23] For Comte in France and Kuno Fischer in Germany, the problem lay not with Bacon but with his followers, who had turned the master into an arch empiricist. According to Comte, while Bacon's own version of the scientific spirit "oscillates between the empirical and the metaphysical," it is his followers who have been and still are "sinking his conception of observation into a sterile empiricism."[24] Fischer, rejecting Goethe's charge that Bacon's method is a "boundless empiricism," says that "[t]his reproach applies to most of those who, at the present day, profess to be followers of Bacon, but not to Bacon himself."[25]

Similarly, Davy's notion that the "precepts of Bacon" bore fruit in the "examples of Newton" also came to be questioned. David Brewster was among the first to deny that Newton's discoveries in mechanics and optics were the result of applying Bacon's methodological principles, referring to "the absurdity of attempting to fetter discovery by any artificial rules."[26] If Brewster's position was extreme in 1831, it had become commonplace, as Richard Yeo notes, by midcentury.[27] According to Augustus De Morgan, the *Novum Organum* "was long held in England—but not until the last century—to be the work which taught Newton and all his successors how to philosophize," but he declared that "[i]f Newton had taken Bacon for his master, not he, but somebody else, would have been Newton."[28]

As the century progressed, criticism of "the common notion of Lord Bacon" became less the occasion for complimentary reinterpretation of Bacon's methods and more the occasion for attack. Men like Herschel and Mill and Whewell believed that science could not have advanced without Bacon and Baconian induction; if the method was inadequate, it was because Bacon's system was incomplete or because Bacon stood at the beginning of a period of scientific reform and therefore lacked copious examples of productive scientific investigations. With the benefit of two hundred years of extraordinary scientific achievement behind them, these writers believed that Bacon could be revised, updated, extended, completed. But such modernization chipped away at the authoritative status of Bacon's specific methodological pronouncements to the point that it did not take a radical shift for Bacon's value to be located in fuzzy statements about his contribution to "the scientific spirit," or even to be repudiated entirely. One studied Bacon not to learn how to do science, but as a historical curiosity; real science was more interesting, more imaginative, and more difficult than Bacon made it out to be. Paradoxically, Baconian methods eventually came to be seen not as an "organ" of discovery, but as an impediment to it.

## FACTS, OBSERVATION, OBJECTIVITY

Essential to the "common notion" of Baconianism was an emphasis on the collection of facts. Induction begins with particulars, and if the investigator is sufficiently patient and industrious, a pattern will eventually arise from the mass of facts. Theories, therefore, emerge from, rather than being imposed on, the data. In Jevons' eyes, such assumptions led to a simpleminded system of scientific accounting:

> Francis Bacon spread abroad the notion that to advance science we must begin by accumulating facts, and then draw from them, by a process of digestion, successive laws of higher and higher generality. In protesting against the false method of the scholastic logicians, he exaggerated a partially true philosophy, until it became as false as that which preceded it. His notion of scientific method was a kind of scientific bookkeeping. Facts were to be indiscriminately gathered from every source, and posted in a ledger, from which would emerge in time a balance of truth. It is difficult to imagine a less likely way of arriving at great discoveries.[29]

Jevons argues that such a process is not really a method at all. In spite of the bookkeeping system, facts are gathered "indiscriminately," and the question of how truth is to emerge from them once they are classified—not to mention how the classification system is to be set up in the first place—is completely begged.

Objections to this bookkeeping system ranged from the practical to the theoretical. Robert Leslie Ellis, one of the editors of the monumental edition of Bacon's *Works* that appeared in 1857–74 and remains standard today, questioned the feasibility of such an approach: how does a scientist know when enough facts, or at least the most important facts, have been collected?[30] Defending Newton's discoveries as the result of genius and imagination, Brewster declared that "[n]othing can be more certain than that a collection of scientific facts are of themselves incapable of leading to discovery, or to the determination of general laws."[31] For Whewell, committed to a philosophy that included necessary truths and fundamental ideas, such an empirical exercise in fact-collecting, generating theory without the mediation of the mind, could only be, as it was for Jevons, a half truth. It testified to a peculiarly English "persuasion" that "the only valuable knowledge is the accumulation of particular facts."[32] But Whewell also believed that the history of science proved that discovery did not take place along the lines Bacon had laid out, a view which was developed in the course of writing his *History* and which led directly to the composition of the *Philosophy*. Coming back to this point in a review of Ellis and Spedding, Whewell argued that "the theories which make the epochs of science do not even grow gradually and regularly out of the accumulation of facts. There are moments when a spring forward is made—when a multitude of known facts acquire a new meaning. . . . Previous to such epochs, the blind heaping up of observed facts can do little or nothing for science."[33]

Whewell's ironic combination of blindness and observation reminds us that the collection of facts occurs principally through the senses, and that the reliability of sense impressions is not unproblematic. "Observations are never perfect. For we observe phenomena by our senses, and measure their relations in time and space; but our senses and our measures are all . . . inaccurate."[34] Bacon had said the same thing, but his nineteenth-century readers had to fight the tradition of the common view, which declared that experience as gathered through the senses was all.[35] Furthermore, while Bacon had offered "helps" for overcoming or at least minimizing the inaccuracy of the senses, his Victorian heirs were compelled to admit that science was characterized by its *distance* from sense impressions. "A broad distinction . . . between scientific knowledge and common knowledge," wrote Herbert Spencer, "is its remoteness from perception."[36] Though caloric and phlogiston had been repudiated, there were other unobservables, from atoms to the ether to magnetic lines of force. How could claims about the nebular hypothesis or the origin of species be verified? Even minerals, once classified on the basis of obvious physical characteristics like color, luster, and hardness, were now differentiated by crystal structure, a feature often indiscernible to the naked eye. For G. H. Lewes, the implications were clear: if science is to be called empirical, then empiricism needs to be redefined. If chief among "inductive principles" is that "we are never to invoke aid from any higher source than Experience," then Experience needs to be liberated from its "unwarrantable restriction" to "sensation." In a passage that suggests the need to reconsider what it meant to be a follower of Comte, Lewes went on to declare that "Science is essentially an ideal construction very far removed from a real transcription of facts. Its most absolute conclusions are formed from abstractions expressing modes of existence which never were, and never could be, real; and are very often at variance with sensible Experience. It not only deals with data that are extra-sensible, but with data avowedly fictitious."[37]

At an even more fundamental level, however, were challenges to the assumption that observation is a neutral activity in which the theoretical predilections of the observer make no difference whatsoever. The observer sees not what he wants to see, but what is. A fact is a fact. Yet when Coleridge mocked the injunction that "you are to observe everything without having reason for so doing," he clearly implied that indiscriminate observation is hopeless, that the scientist must have some provisional theory to aid him first in deciding what to observe and then in determining how to interpret it. Observation, as the expression goes, is theory-laden—though in the nineteenth century the tentative development of this "truism" constituted a radical break from the orthodoxy of naive Baconianism.

Consider William Whewell. For Whewell, facts and theories were inseparable. Science always requires both facts and ideas, clearly expressed and bound together: "our knowledge consists in applying Ideas to Facts; . . . the conditions of real knowledge are that the ideas be distinct and appropriate, and exactly applied to clear and certain facts." Without facts, we have empty speculations. Without ideas, we have chaos. Even what we call facts are often cases where we are not aware that we are interpreting nature, where earlier inferences have been so well confirmed that we are unaware of their presence.[38] The implication is that ultimately there is no such thing as pure facts, but such an implication would just as clearly threaten the very foundations of science's authority, its access to knowledge that is true and permanent. More often, then, despite his insistence on the inseparability of fact and theory and his reliance on necessary truths, Whewell assumes that at some level there is a ground of fact on which theories are imposed: "the peculiar import" of induction is "Conception *superinduced* upon the Facts."[39]

This tension is evident in Whewell's project as a whole. Whewell denies that the goal of a philosophy of science is to provide, as Bacon tried to provide, a strict method of discovery. Rather, we study the history of science to classify and analyze, to see how discoveries *have* occurred, not how they *must* occur. At best, such a study offers a loose guide for future researchers.[40] Yet for all his disclaimers, Whewell is offering an incredibly thorough and specific philosophy of science, the authority of which is based on historical "facts." The massive *Philosophy* is possible only after the equally massive foundation of fact in the *History*. And it is the factual basis of the *History*, Whewell claims, that makes his *Philosophy* complete, objective, and true:

> With so much real historical fact before us, we may hope to avoid such views of the processes of the human mind as are too partial and limited, or too vague and loose, or too abstract and unsubstantial, to represent fitly the real forms of discovery and of truth. . . . [A] Philosophy of the Sciences ought . . . to be founded, not upon conjecture, but upon an examination of many instances;—should not consist of a few vague and unconnected maxims, . . . but should form a system of which every part has been repeatedly confirmed and verified.[41]

Whewell's philosophy of induction is itself inductive, yet it relies on some of the very claims about the common view of Baconianism that the body of his work most thoroughly challenges.

The methodological writings of the century are in almost unanimous agreement that facts, observation, experiment, and even the total concept of induction itself are all theory-laden. Comte declared that "if it is true that every theory must be based upon observed facts, it is equally true that facts can not be observed without the guide of some theory. Without such guidance, our facts would be desultory and useless; we could not retain them: for the most part we could not even perceive them."[42] Mill's position on observation was the same as Whewell's on facts: what we call observation is almost never pure observation, but at least partly inference.[43] Ellis echoed this, proclaiming that "the observations necessary in order to the recognition of these facts would never have been made except under the guidance of some preconceived ideas as to the subject of observation."[44] Manipulated observation—experiment—also presupposes theory. In his scathing attack on Bacon, the great German chemist Justus von Liebig ridiculed Baconian experiments as "practiced without knowing what we are about. They are to be compared to acts without a motive, and their result is therefore without end or gain." Liebig agreed that experiments are necessary if we are to correct for the inaccuracy of the senses, but he argued that the setting up of an experiment cannot occur without the setting up of a theory: "in all science investigation is deductive, or *a priori*. The experiment is but the aid to the process of thought, . . . and the thought, the idea, must always precede it—necessarily precede it."[45]

If theory must precede, or at least go hand in hand with, observation and experiment, then it would seem that the scientist runs the risk of determining what he sees and how he interprets it before the process begins. Yet this was, generally speaking, not a focus of concern. Instead, writers about scientific method worried, in keeping with Bacon's doctrine of idols, that an investigator might become so enamored with his own theory that he would privilege it over other equally valid ones, or that he would willfully misinterpret or ignore inconvenient data. The need for theory did not threaten the validity of the data itself. Just because an observer has to form a theory before he can collect facts does not mean that he cannot be objective in assessing the fit between fact and theory. The authority and integrity of science demand at least this much effacement of the scientist's prejudices.

Although his *Preliminary Discourse on the Study of Natural Philosophy* goes on to develop the importance of hypothesis, John Herschel opens his discussion of method with the need for beginning any investigation with a mental tabula rasa as well as for ending it with an unbiased review of the facts:

> [I]n the study of nature and its laws, we ought at once to make up our minds to dismiss as idle prejudices, or at least suspend as premature, any preconceived notion of what might or what ought to be the order of nature . . . and content ourselves with observing, as a plain matter of fact, what is. To experience we refer, as the only ground of all physical inquiry. But before experience itself can be used with advantage, there is one preliminary step to make. . . . [I]t is the absolute dismissal and clearing the mind of all prejudice . . . and the determination to stand and fall by the result of a direct appeal to facts in the first instance, and of strict logical deduction from them afterwards.[46]

"Science," he wrote in a review of Whewell's *Philosophy*, "is essentially abstract, passionless, disinterested. Results are to be accepted for their truth alone."[47] Techniques like the replication of experiments, or having an unknowledgeable laboratory assistant make observations, can be employed to ensure prejudice is eliminated or minimized from the collection of data, but the scientist is on his honor to extend his disinterestedness to the interpretation of results. "[T]he habit of forming a judgment upon these facts unbiased by personal feeling is characteristic of what may be termed the scientific frame of mind," wrote Karl Pearson near the end of the century.[48]

The disinterested judgment of truth naturally provoked comparisons to a court of law. In John Theodore Merz's words, "[t]he scientific mind should acquire . . . an attitude as dispassionate and as evenly balanced as that of a judge to whose care the most momentous issues . . . are intrusted."[49] Yet challenges to such an analogy were not wanting. Though praising Bacon's emphasis on impartial observation, Jevons pointed out that biased interest in verifying a pet theory often produced good observation.[50] De Morgan drew directly on the comparison of natural and civil law, pointing out that from the Baconian perspective the scientist is like the judge, waiting to hear all the facts and then applying a set of rules to come to a decision. But, continued De Morgan, the analogy does not really apply, for the scientist, interpreting and writing the laws of nature, simply cannot be disinterested: "Now the truth is, that the physical philosopher has frequently to conceive law which never was in previous thought—to educe the unknown, not to choose from the known. . . . Bacon seems to us to think that the philosopher is a judge who has to choose, upon ascertained facts, which of known statutes is to rule the decision: he appears to us more like a person who is to write the statute book."[51] The myth of objectivity is not rejected to the same degree that the myth of the separation of fact and theory is, but here, too, method brings the individual scientist, with his quirks and prejudices and limitations, back toward the center of the scientific enterprise—and in some cases celebrates his inability to be objective, neutral, and dispassionate.

## GRADUALNESS, ELIMINATION, MECHANISM

In articulating his version of induction, Bacon distinguished it from simple enumeration on the one hand, rapid movement from a few particulars to general truths on the other. Collecting facts did not mean collecting all facts, but employing selection and discrimination. Flying off to hasty generalizations on the basis of insufficient evidence was equally inadequate, however, for it rendered useless all the deductions made from these tenuous general principles. Instead, Bacon urged, science must proceed gradually, collecting sufficient numbers of important facts, eliminating possible explanations until the true one remains, and then moving from that induction to a new one of wider generality. By traveling rung by rung up the inductive ladder, the scientist eventually reaches the highest truths, truths of which he can be certain because he has been certain in climbing each rung. From these general truths he can then carry his investigations into new areas, confident that his deductions are also valid.

If Bacon's distinction between his version of induction and simple enumeration had become blurred in the intervening centuries, his insistence on gradual movement rather than hasty generalization obviously had not. Warnings were often sounded in the early years of the century about this dangerous tendency. Brougham launched his attack of Thomas Young's wave theory of light with the complaint that "a most unaccountable predilection for vague hypothesis [is] daily gaining ground"—a predilection more worrisome as Young's lecture was delivered before the Royal Society.[52] Davy, no stranger to imaginative leaps himself, nonetheless warned against "hasty generalizations" and "premature development of theory."[53] Herschel lamented in an address to the British Association for the Advancement of Science the "propensity which is beginning to prevail widely, and, I fear, balefully, over large departments of our philosophy, the propensity to crude and hasty generalization."[54]

In spite of this concern, most writers, including those who identified and supported Bacon's call for restraint, recognized that discovery demanded imaginative flights. In the *Philosophy,* Whewell says that Bacon's emphasis on the inductive process as gradual and successive "exactly represents" the "general structure of the soundest and most comprehensive physical theories," and that anyone who has studied the "progress of science" can attest to this. Yet earlier in the same work he declares that "[a]t each step in the progress of science are needed invention, sagacity, genius;—elements which no Art can give."[55] In the *History,* "real discoveries" are said to be "mixed with baseless assumptions," "profound sagacity" combined with "fanciful conjectures"—"not rarely, or in peculiar instances, but commonly, and in most cases."[56] Herschel, though careful to deny that in the formation of theories we are "abandoned to the unrestrained exercise of the imagination," nonetheless admits that strict induction rarely occurs in practice. In "[t]he study of nature, we must not . . . be too scrupulous as to how we reach to a knowledge of such general facts, provided only we verify them carefully when once detected."[57]

If Whewell and Herschel were inconsistent, however, others were not. Brewster marshaled examples from the history of science to prove that the scientist, impatient of rules and the drudgery of gradual ascent, practices "the very reverse of the method of induction" by letting his imagination form theories at will.[58] By the time John Nichol prepared his work on Bacon's life and philosophy for the Philosophical Classics for English Readers series in 1907, there was no question that Bacon "ignored the necessity for a divining power in science, and so underestimated the value of the Deductive Method."[59]

Bacon was sure that once a list of possible causes had been prepared, the systematic elimination of those causes would eventually yield one certain cause—his method had no place for any "divining power." Yet it is difficult to find much support for this view in the nineteenth century, even among those less hostile than Brewster to the drudgery of Baconian elimination. In his revision of Bacon's inductive rules, Mill specifically calls two of his own methods of experimental inquiry methods of elimination, likening them to Bacon's process of exclusion. Yet these methods are at best half the story, for "Bacon's greatest merit cannot . . . consist, as we are so often told that it did, in exploding the vicious method pursued by the ancients of flying to the highest generalizations first, and deducing the middle principles from them; since this is neither a vicious nor an exploded, but the universally accredited, method of modern science."[60]

If "flying to the highest generalizations first" is "the universally accredited method of modern science," there is a good practical reason for it: assembling a complete roster of possible explanations and systematically testing not only each individual explanation but also various combinations, is endless. Moreover, there must always be doubt that we have assembled a truly complete list in the first place. Hume had argued in the eighteenth century that induction can never generate absolute certainty because no number of individual cases, however great, can justify the assumption that the next case will not contradict the general law. In the nineteenth century, Jevons sees that this doubt must inevitably color the process of exclusion as well: "The logical rule is—Try all possible combinations; but this being impracticable, the experimentalist necessarily abandons strict logical method, and trusts his own insight. Analysis . . . gives some assistance. . . . But we are now entirely in the region of probability."[61] Insight is essential if scientific method is not to be mired in drudgery, but in no case can certainty be achieved. Jevons imports probability theory into the study of scientific method because "[n]o inductive conclusions are more than probable," and therefore "the logical value of every inductive result must be determined consciously or unconsciously."[62] Furthermore, this strict logical method of elimination is too mechanical. It again excludes the human imagination from the process of science. Whewell rejects elimination because it is a negative process which wrongly assumes that rejecting what is false will leave us with what is true. Rather, he says, "nothing but a peculiar inventive talent could supply that which was thus not contained in the facts, and yet was needed for the discovery."[63]

In Aphorisms 61 and 122 of the first book of the *Novum Organum* Bacon insisted that it was precisely the lack of a need for any special "inventive talent" in his method that made scientific discovery possible for anyone. In this Protestant democracy of the intellect, every person has the ability to interpret the Book of Nature for himself. In the view of Ellis, Bacon's "mechanical mode of procedure" makes possible "the two great features of the Baconian method": "absolute certainty" and the rendering "all men equally capable, or nearly so, of attaining truth." Central to the mechanical mode of procedure is the process of exclusion, for "this method of exclusion requires only an attentive consideration of each 'instantia,' in order first to analyse it into its simple natures, and secondly to see which of the latter are to be excluded—processes which require no higher faculties than ordinary acuteness and patient diligence. There is clearly no room in this mechanical procedure for the display of subtlety or of inventive genius."[64]

In his review of Ellis and Spedding, Whewell endorsed their criticism of this mechanical procedure, a criticism of Bacon that he had been offering for two decades. It is wrong, Whewell wrote, to believe that "discovering in science can be made without any special inventive aptitude; that the facts may be collected, and then treated in some regular way, so that the scientific truth shall emerge in virtue of the method alone; all men being alike, or nearly alike, able to perfect the operation when the right method is followed." Bacon, like all who "try to devise technical methods of extracting science from facts," overlooks the "great truth" that "the process of discovery necessarily involves invention—mind—genius." Bacon's method is "pervaded" by the "errour" of "supposing that to be done by method which must be done by mind; that to be done by rule which must be done by a flight beyond rule; that to be mainly negative which is eminently positive; that to depend on other men which must depend of the discoverer himself; . . . that to be a work of mere labour which must be also a work of genius."[65]

Others argued that Bacon's method, if not strictly mechanical, was at least the systematization of common sense. By blurring just what was meant both by "systematization" and by "common sense," this view had a certain appeal, for it retained concepts of rigor and intellectual equality while excluding a rigid mechanism of discovery and allowing some room for inventiveness. Criticisms like Macaulay's—"[t]he inductive method has been practised ever since the beginning of the world by every human being" (including, in Macaulay's deliberately homely example, the case of the ignorant peasant who determines that kidney pie has caused his upset stomach)—could be met with the claim that the methods of science went well beyond common sense.[66] Herschel, for example, provided many examples where "ordinary practical reasoning" was of no use.[67] Yet science as common sense was a valuable rhetorical strategy for popularization, for showing that science, however bizarre or disconcerting its results, was not inaccessible, not far removed from everyday events and everyday minds. Huxley was perhaps the greatest exploiter of this rhetoric, yet the notions he tried to popularize often gave the lie to the claim that they were the result of common sense.[68] And when Sherlock Holmes uses the phrase "systematized common sense" to describe his own methods, we cannot help but wonder if these appeals to common sense are not at least somewhat disingenuous.[69]

## HYPOTHESIS AND IMAGINATION

In the effort to make scientific method more imaginative and creative, the most crucial issue was the status of hypothesis, for it was here that the illustrious authority of Newton was seen to merge with that of Bacon. Bacon, so the story went, rejected hypotheses. Newton, the celebrated Newton, discoverer of the theory of universal gravitation, confirmed this rejection. "Hypotheses non fingo," Newton had said. I make or frame, as it was variously translated in the eighteenth century, no hypotheses.[70]

The power of this position is invoked by Brougham in his attack on Young's "vague hypothesis" of the wave theory of light in 1803. At the time, the corpuscular theory reigned, in large measure due to the authority of Newton himself, so Brougham's vituperative assault was designed to defend both the matter and the method of the master. Turning Bacon and Newton into Christ figures, Brougham complained that hypotheses like Young's "renew all those wild phantoms of the imagination which Bacon and Newton put to flight from her temple. We wish to recall philosophers to . . . strict and severe methods of investigation. . . ." As a "mere theory," Young's hypothesis was "destitute of all merit." It was intellectual sodomy, the product of an "unmanly and unfruitful pleasure of a boyish and prurient imagination, or the gratification of a corrupted and depraved appetite."[71]

The challenge facing those who argued for the importance of hypothesis lay not so much in repudiating this powerful element of the Baconian-Newtonian legacy as in contextualizing and reinterpreting it. One would prefer not to reject Bacon, but one definitely couldn't go without Newton, the undisputed giant—and an English giant, moreover—of physical science. So in defending the methodological importance of hypotheses, certain strategies evolved. Bacon and Newton could be safely criticized for underestimating the importance of hypotheses so long as it was pointed out that both of them did in fact employ hypotheses, or that their criticism was directed only at certain types of hypotheses.

In Bacon's case, commentary focused on his application of his new method to the example of heat. After employing the process of exclusion, Bacon still had no answer to the question "What is the nature of heat?" He was therefore forced to offer what amounted to a preliminary hypothesis, in his terminology an "Indulgence of the Understanding" or "Commencement of Interpretation" or "First Vintage."[72] This failure of Bacon's method was interpreted critically by Brewster as a "memorable instance" of the "absurdity" of Bacon's rules, while for Fowler it was a "happy" inconsistency that inadvertently showed both the importance and the proper use of hypotheses.[73] But whatever the interpretation, it was agreed that Bacon's own example established the centrality of hypotheses in scientific method.

Newton's case was similar, though Newton was generally converted, especially after midcentury, into the champion of the hypothetico-deductive method. Whewell blamed Bacon and Newton for the fact that hypotheses had "fallen into a kind of obloquy" as "dangerous temptations and fallacious guides" when, on the contrary, they are not only useful but indispensable: "A facility in devising hypotheses . . . is so far from being a fault in the intellectual character of a discoverer, that it is, in truth, a faculty indispensable to his task. It is . . . much better that he should be too ready in contriving, too eager in pursuing systems . . . than that he should be barren of such inventions. . . ." If we read Newton carefully, we find that he had this facility, and we conclude that his example justifies the use of such conjectures. There are limitations—hypotheses should be clearly expressed, grounded in fact, and carefully tested—but such a procedure "belongs to the *rule* of the genius of discovery, rather than (as has often been taught in modern times) to the exception." Indeed, so long as the experimenter does not cling to his hypothesis in the face of contradictory evidence, "to try wrong guesses is, with most persons, the only way to hit upon right ones."[74]

Herschel endorses Whewell, saying that the hypotheses "proscribed" by Newton "in his celebrated 'hypotheses non fingo'" were those that were "loose and incapable of being exactly tested."[75] Whewell, like others before him, had in fact been more specific: the hypotheses Newton had in mind were those of Descartes, whose method was described as "rash and illicit" and "perverse."[76] While Lewes also regrets that "the weight of [Newton's] authority has pressed Hypothesis into the mire" and turned it into "the pariah of research," he, too, affirms that though "Newton's assertion, *Hypotheses non fingo*—I make no hypotheses—has been incessantly repeated by men who fancy themselves Baconian thinkers," Newton himself "gives it no countenance." Instead, hypotheses are an "imaginative arch," a scaffolding on which the temple of science is constructed. "The wildest flights of Imagination" are permitted so long as they ultimately submit to verification.[77] In similar fashion Mill declared that the function of hypotheses "must be reckoned absolutely indispensable to science. When Newton said, 'Hypotheses non fingo,' he did not mean that he deprived himself of the facilities of investigation afforded by assuming in the first instance what he hoped ultimately to prove. Without such assumptions, science could never have attained its present state."[78]

It begins to be clear that maintaining a line of unbroken methodological continuity from Bacon to Newton to contemporary nineteenth-century science was extraordinarily difficult. If hypotheses are a cornerstone of scientific method, then Newton almost inevitably replaces Bacon as a methodological titan, in part because it is easier to explain away hypotheses non fingo than the virtual entirety of Baconian induction. As master of Trinity College, Whewell erected in the antechapel a statue of Bacon to join the one of Newton, but his "renovation" of the *Novum Organum* reveals the strain of trying to combine the methodologies of the two men. Whewell could be a traditionalist, calling Newton's theory of universal gravitation "even now, by far the best example of inductive reasoning" and the "pointed actual exemplification" that Bacon's work lacked. Elsewhere, however, Newton's discovery epitomized for Whewell that indispensable element of genius in the formation of hypotheses that scientific method could not do without. On one occasion, Whewell even says that Newton has the air of one who works in a burst of inspiration, "as an irresistible and almost supernatural hero of a philosophical romance."[79]

Thomas Fowler, who gave more favorable assessments of Bacon's character and method than most late-century writers, also tried to maintain the connection between Bacon and Newton: "'Hypotheses non fingo' was a maxim which Newton inherited directly from the teachings of Bacon. And, though the reaction against hypotheses was carried much too far, and though Bacon's utterances on the subject, to be serviceable at the present time, require much rectification, the warning . . . in his own time, was sorely needed."[80] Yet Fowler's position seems to sunder contemporary science from the very tradition that he is so intent on preserving. He aligns himself with Mill, whom he sees as articulating a stricter and more Baconian form of induction, over those like Whewell and Jevons, whom he sees as endorsing a looser method built around hypotheses. But Fowler's somewhat one-sided reading of Mill and his preference for inductive over hypothetico-deductive methods are undermined by his criticisms of Bacon for depreciating hypotheses and the imagination. As the qualifications pile up and Bacon's influence becomes vaguer and vaguer, one wonders how much "rectification" Bacon's utterances can stand.

By Fowler's day, most thinkers had drawn the line. What Fowler called one of the versions of induction others identified as a different method, associated with the practice if not the overt precepts of Newton. In his preference for Mill over Whewell, Fowler magnified the importance of relatively minor differences that had already been exaggerated by the running argument between the two men about certain aspects of induction, an argument carried on in the successive editions of Mill's *Logic* and Whewell's *Philosophy.* Mill was more interested in the logical validity, Whewell in the historical reality, of induction. Thus for Whewell, whose primary focus was discovery, hypotheses which were verified and which explained relevant facts could be considered true and used as the basis for future speculation and investigation. For Mill, however, the key issue was proof. It is often possible to construct several different hypotheses that account for the same group of facts; therefore, inductive truth is obtained only when one hypothesis is verified *and* the others are shown to be false. Perhaps sensing that their public debate was obscuring important areas of agreement, Mill pointed out that both he and Whewell "have maintained against the purely empirical school" that it is "seldom possible . . . to go much beyond the initial steps [of an induction] without calling in the instrument of Deduction, and the temporary aid of hypotheses."[81] Mill drew heavily on Herschel's *Preliminary Discourse* and like Herschel was opposed to Whewell's idealism, but Herschel himself claimed that Mill's and Whewell's views on induction were "identical" in the "most essential features" when "carefully examined."[82]

For Comte, this shift from induction to hypothesis and deduction had both a logical and a chronological basis. Neither pure induction nor pure deduction would help us "if we did not begin by anticipating the results, by making a provisional supposition, altogether conjectural in the first instance, with respect to some of the very notions which are the object of inquiry. Hence the necessary introduction of hypotheses into natural philosophy." If the observation of facts provides a beginning for method, hypotheses are still essential for getting us to the general laws from which deductions can be made. As a science's general laws are established, it becomes more and more deductive, able to predict phenomena rather than relying primarily on direct observation: "We have . . . sanctioned . . . the now popular maxim of Bacon, that observed facts are the only basis of sound speculation. . . . On the other hand, we have repudiated the practice of reducing science to an accumulation of desultory facts. . . . Besides that sound theoretical indications are necessary to control and guide observation, the positive spirit is forever enlarging the logical province at the expense of the experimental, by substituting the prevision of phenomena more and more for the direct exploration of them."[83]

Mill's position on this issue is essentially the same as Comte's. Whereas Bacon rejected the deductive method because it had been abused by those employing arbitrary premises, the maturation of science now makes possible a sound deductive method with inductively established premises. The deductive method is "destined irrevocably to predominate in the course of scientific investigation from this time forward," for "[a] revolution is peaceably and progressively effecting itself in philosophy, the reverse of that to which Bacon has attached his name. That great man changed the method of the sciences from deductive to experimental, and it is now rapidly reverting from experimental to deductive."[84]

While acknowledging that "Bacon was not wholly unaware of the value of hypothetical anticipation," Jevons also contended that Newton's method of hypothesis and deduction, not Baconian induction, was the method of current science. Hypotheses non fingo, says Jevons, "bears the appearance of irony," not only for us, who can see that deductive reasoning was "wholly predominant" in Newton's work, but for Newton's contemporaries as well. So aggressive is Jevons on Newton's behalf that he displaces Bacon from the methodological pantheon in favor of Newton. In a section headed "The Newtonian Method, the True Organum," Jevons calls the *Principia* "the true *Novum Organum* of scientific method." Both Bacon and Newton founded their methods on observation and experience, but Newton seized on the "true method" of carefully formulated, carefully verified hypotheses. Consequently, "[i]t is a great mistake to say that modern science is the result of the Baconian philosophy; it is the Newtonian philosophy and the Newtonian method which have led to all the great triumphs of physical science. . . ."[85]

Other endorsements of the hypothetico-deductive method at about this time are plentiful. Jevons used Tyndall as an example of a working scientist and methodological theorist who saw the value of hypothesis, for Tyndall told the British Association in 1868 that "the investigator proceeds by combining intuition and verification. . . . [H]e guesses, and checks his guess; he conjectures, and confirms or explodes his conjecture. . . . The force of intellectual penetration . . . is not, as some seem to think, dependent upon method, but upon the genius of the investigator."[86] The ninth edition (1878) of the *Encyclopaedia Britannica* declared in its article on Bacon that "[t]he true scientific procedure is by hypothesis followed up and tested by verification; the most powerful instrument is the deductive method, which Bacon can hardly be said to have recognized." Writing in 1872, De Morgan claimed that "[m]odern discoveries have not been made by large collections of facts, with subsequent discussion, separation, and resulting deduction of a truth thus rendered perceptible." Rather, a "few facts have suggested an hypothesis" which "must have been started, not by rule, but by that sagacity of which no description can be given. . . ."[87]

Looking back at Herschel's *Preliminary Discourse*, both Jevons and De Morgan found that in spite of Herschel's evident respect for Bacon, his incorporation of hypotheses brought him closer to Newton's practice than to Bacon's rules. The reason for this, according to De Morgan, was that Herschel had vividly before him the example of his father, the astronomer William Herschel, "whose processes would have been held by Bacon to have been vague, insufficient, compounded of chance work and sagacity, and too meagre of facts to deserve the name of induction."[88] De Morgan's speculation receives confirmation from Herschel himself, who wrote to Whewell, "I remember it was a saying often in my father's mouth 'Hypotheses fingo,' in reference to Newton's 'Hypotheses *non* fingo.' "[89] In the new interpretation of the history of British science, Newton had found the via media between Descartes's "loose" hypotheses and Bacon's depreciation of all hypotheses. If eighteenth-century scientists had too often misinterpreted hypotheses non fingo, they had at least had the good sense in practice, like William Herschel, to employ hypotheses nonetheless.

No matter how much or how little Bacon was considered in accord with the "real" methods of science, all commentators endeavored to cast science as a more imaginative enterprise. The problem with Bacon's method, said Whewell, is that it supposes "that to be mere prose which must have a dash of poetry."[90] Science requires intuition, genius, invention, imagination; it cannot succeed by excluding such things. Yet science is different from poetry, and nineteenth-century scientists no more wanted to validate their discipline by comparing it to poetry than poets wanted to validate their discipline by comparing it to science. Thus almost any celebration of the scientific imagination is accompanied by a qualification or a caveat—the "prose" of science contains only a "dash" of poetry, the scientific imagination must always be controlled, brought into "rigid contact with facts" (this phrase is also Whewell's), and verified.

But what, it must be asked, do these limits amount to? In the same passage where conjectures are to be brought into rigid contact with facts, Whewell also says that "baseless assumptions," "fanciful conjecture," and "wrong guesses" are not only common but "the only way" to obtain truth. When Herschel urged the controlled use of the imagination, he relied on metaphoric language: "The liberty of speculation which we possess in the domains of theory is not like the wild licence of the slave broke loose from his fetters, but rather like that of the freeman who has learned the lessons of self-restraint in the school of just subordination."[91] Herschel's simile rings with socially conservative language suggesting that unrestrained imagination is a disruptive and anarchic force in the hierarchical and possibly authoritarian world of science, yet the association of speculative liberty with an escaped slave cannot help but imply that speculation had been wrongly fettered. Similarly, in a highly favorable review of the works of the Belgian astronomer Quetelet, Herschel nonetheless complains that the fictional epistolary format of Quetelet's books distracts the reader's attention from their scientific matter. Herschel says his objection is "against all such artifices of communication as letters—dialogues—catechisms, &c., if the subject be a scientific one and the object of the work didactic." Yet Herschel immediately illustrates his own didactic point with an artifice of his own: "They [the artifices] are like pebbles in the bed of a stream, which may make it sparkle and please the eye and ear when the thought is but loosely engaged. But the welling waters of scientific lore should be clear, glassy, and unrippled, offering their inmost depths to a quiet and contemplative gaze, and neither distracting by murmurs nor dazzling by irregular reflections."[92]

John Tyndall, who, along with Huxley, represented to many the apex of scientific materialism, delivered addresses in 1868 and 1870 on the "Scientific Limit of the Imagination" and the "Scientific Use of the Imagination." In the latter, Tyndall insisted that scientists must permit their imaginations to rise above the physical limitations of vision. In the former, he offered an equally eloquent defense of the imagination. Scientific materialism, he argued, was only half of the truth: materialist explanations don't really explain, in the sense that they leave unresolved the mysteries of the human mind, of human origins, of humanity's relation to the universe. Though writing about the limits of the imagination in science, Tyndall defined "the vocation of the true experimentalist" as "the continued exercise of spiritual insight, and its incessant correction and realization. His experiments constitute a body, of which his purified intuitions are, as it were, the soul."[93] So powerful was Tyndall's formulation that by the end of the decade the "scientific use of the imagination" had become a catchphrase, used by the *Encyclopaedia Britannica* to capture all that was lacking in Baconian induction. The "power of framing hypotheses" is vital to the "true method" of science, a method in which the mind "anticipates nature, and moulds the experience received by it in accordance with its own constructive ideas or conceptions." In this un-Baconian notion of scientific method (science, Bacon insisted, must interpret nature, not anticipate her), there is "room for the scientific use of the imagination, and for the creative flashes of genius."

Both Tyndall and Liebig asserted the interconnectedness and equality of the poetic and scientific temperaments—in Liebig's words, "the mental faculty which constitutes the poet and the artist is the same as that whence discoveries and progress in science spring."[94] W. B. Carpenter, in his presidential address to the British Association in 1872, claimed that the "Scientific interpretation of Nature" only seems less individualistic than that of the artist or poet: "it can be shown to be no less true of the Scientific conception of Nature, than it is of the Artistic or the Poetic, that it is a *representation framed by the mind itself* out of the materials supplied by the impressions which external objects make upon the senses."[95] Comte, despite arguing that the movement from the theological to the metaphysical to the positive was characterized by "the steady subordination of the imagination to observation," lamented the "anti-aesthetic character" that "the sway of the mathematical spirit" had exercised on positive philosophy. In the future, he predicted, the "guiding spirit" of positivism will be sociology rather than mathematics, and this new guiding philosophy will be "more favourable to Art" and will "establish an irreversible agreement between the aesthetic and scientific points of view."[96]

This positivist agreement between the aesthetic and the scientific was carried out in significant measure in England by G. H. Lewes. In his 1853 exposition of Comte, Lewes reiterated that positive philosophy "is characterized by that necessary and permanent subordination of the imagination to observation which specially constitutes the scientific spirit. . . ."[97] Two years later, however, in his *Life of Goethe*, Lewes argued that Goethe, like the Realist in philosophy or science who moved from Nature upward rather than from Ideas downward, controls "the errant facility of imagination" with "an imperious desire for reality." Although the pretensions of the Realist are less lofty than those of the Idealist, the Realist "achieves loftier results."[98] The controlled imagination, whether in science or in art, ultimately ascends to higher truths than it does when unfettered from the start. Moreover, such truths really are stranger than fiction; the scientific imagination out-imagines even the "wayward caprices" of its unrestrained poetic counterpart: "the truth is that Science mounts on the wings of Imagination into regions of the invisible and impalpable, peopling these regions with Fictions more remote from fact than the fantasies of the Arabian Nights are from the daily occurrences in Oxford Street."[99] Imagination ceases to be subordinate and becomes instead the vehicle of scientific truth. The fantasies of the Arabian Nights are on a par with the constructions of science—with cells and molecules, with light and the ether, with non-Euclidean geometry and the origins of language.

For some, the incorporation of imagination into scientific method meant that science could lay claim to a higher status than poetry, which became "merely" imaginative in the way that science had been "merely" empirical. According to this view, science combines the subjective and the objective, observer and observed, personal revelation and general truth. In the 1802 preface to the *Lyrical Ballads*, Wordsworth had envisioned the possibility of such a fusion, the day when the poet "will be ready to follow the steps of the man of science. . . ." Before the sentence is over, however, the poet is no longer following the man of science but is "at his side." By the end of the passage, science's "transfiguration" into "a form of flesh and blood" will require the aid of the poet, who is already *both* a "divine spirit" and a "genuine inmate of the household of men."[100] Ninety years later, Karl Pearson turns the tables on Wordsworth. According to Pearson, the poet is a valued member of the community, but his value will increase as he gains insight from science. The "sublime language" of the poet "will not satisfy our aesthetic judgment, our idea of harmony and beauty, like the few facts which the scientist may venture to tell us. . . . Our aesthetic judgment demands harmony between the representation and represented, and in this sense science is often more artistic than modern art." Referring to Wordsworth's preface, Pearson remarks that the aesthetic judgment and the scientific judgment are "exactly parallel," but then goes on to elevate science in the way that Wordsworth earlier elevated poetry: "the scientific interpretation of phenomena, the scientific account of the universe, is therefore the only one which can permanently satisfy the aesthetic judgment, for it is the only one which can never be entirely contradicted by our observation and experience."[101]

Far from repudiating the imagination, then, nineteenth-century scientists and philosophers saw it as an indispensable component of scientific method. Hypotheses could not be formed, analogies could not be recognized, inferences could not be made without it. Scientists were poets in their own right. In Tyndall's words, "[t]he world embraces not only a Newton, but a Shakespeare—not only a Boyle, but a Raphael—not only a Kant, but a Beethoven—not only a Darwin, but a Carlyle. Not in each of these, but in all, is human nature whole. They are not opposed, but supplementary—not mutually exclusive, but reconcilable."[102] Such claims were not unanimously supported within the scientific community, and when developed in the fashion of Pearson they could generate anxiety among the artists. But in general terms they are representative of a broad movement that saw the man of science growing closer to the poet, that consciously conceived of scientific method as a valuably and inescapably imaginative process, and that even spoke of science as a literary discipline complete with its own "dash of poetry."[103]

## BACON, NEWTON, AND THE STATE OF SCIENCE

The debate over methodology had ramifications for the status of the historical figure involved. As Whewell had seen so clearly, the development of a philosophy of the inductive sciences could not take place apart from a history of the inductive sciences. In Bacon's case, the result of the methodological debate was a deep ambivalence toward his historical position, present if not affirmed even in those most supportive of induction. Well after his methodology could be dismissed in its details, Bacon remained a powerful figure in the British scientific tradition—though his name had to be invoked in rather different ways. To demonstrate this ambivalence, I want to consider first the example of Whewell, and then the various nineteenth-century appropriations of Cowley's image of Bacon as Moses.[104]

In the prefaces to both the *History* and the *Philosophy,* Whewell refers to the need to revise and update Bacon's methods of discovery. Himself a member of "the College of Bacon and Newton," Whewell locates his revision in a tradition of reform beginning with Bacon's own reform of Aristotle. He argues that with the benefit of two extra centuries of scientific discovery, he will develop "a *Philosophy of Science,* . . . the

*New Organ* of Bacon, *renovated* according to our advanced intellectual position and office."[105] Whewell even goes so far as to imitate the format of the *Novum Organum* by reducing his text to aphorisms, and when, in later editions, he split the text into separately titled volumes, he called one of them the *Novum Organon Renovatum*. In spite of such homage, however, we quickly see that this work, "intended as an application of the plan of Bacon's Novum Organum to the present condition of Physical Science," performs considerable renovation indeed:

> Bacon only divined how sciences might be constructed; we can trace, in their history, how their construction has taken place. . . . And with respect to the methods by which Science is to be promoted,—the structure and operation of the *Organ* by which truth is to be collected from nature,—we know that, though Bacon's general maxims still guide and animate philosophical enquirers, yet that his views, in their detail, have all turned out inapplicable: the technical parts of his method failed in his hands, and are forgotten among the cultivators of science. It cannot be an unfit task, at the present day, to endeavour to extract from the actual past progress of science, the elements of a more effectual and substantial Method of Discovery.[106]

If the details of Bacon's views are inapplicable, if the technical parts of his method have failed and are forgotten, one wonders how any renovation can salvage them. Surely what Whewell offers is, by his own admission, an organ very different from Bacon's, yet formulated in language that simultaneously pays tribute to both the power and the paucity of Bacon's method.

This ambivalence is evident in all its complexity in Whewell's descriptions of his "Inductive Tables," charts designed to show in shorthand form the history of scientific induction in different fields.[107] They are arranged as inverted genealogical tables (Whewell himself makes this comparison), with inductive truths converging downward into successively more general laws. Whewell also calls these tables "Inductive Trees." Every junction of two branches represents an induction; the base of the trunk represents a science's ultimate generalization. Such an organic model suggests a certain inevitability in the growth of science, a tendency toward increasingly general laws, and an interconnection of inductions. Furthermore, each element on the tree reflects Whewell's commitment to the inseparability of fact and theory, for each entry is both a fact (relative to the entry above it) and a theory (relative to the entry below it).

Yet Whewell's trees also have their drawbacks. They grow backward from their upper branches to the ground. The union of two branches, Whewell admits, is "very inadequate to convey the true state of the case," for "in Induction, . . . besides a mere collection of particulars, there is always a new conception, a principle of connexion and unity, supplied by the mind, and superinduced upon the particulars." Though the trees capture the fact/theory fusion and the *historical* process of gradual development, they cannot capture the essence of method itself. Whewell speaks of the processes of deduction and verification as traveling over the same inductive path but in the opposite direction, from trunk to branches. But his emphasis on the importance of inference and hypotheses and imaginative leaps makes it impossible for the trees to stand as models of scientific method. Indeed, Whewell says here that *Deduction* "descends steadily and methodically, step by step," while *Induction* "mounts by a leap which is out of the reach of method. She bounds to the top of the stair at once." Obviously aware of these drawbacks, Whewell uses a series of verbal metaphors to elucidate his visual one, but these new comparisons expose as many tensions as they reconcile.

Whewell first tries the genealogical table analogy, eventually extending it by saying that the addition of the discoverers' names to their discoveries makes for a "Genealogical Tree of scientific nobility." This creation of a social hierarchy is attractive to Whewell because it is in accord with the need for genius in scientific discovery. But such a view makes science a bit too exclusive and privileged—in the previous decade, after all, Whewell had supported Herschel's candidacy for president of the Royal Society, an election in which reform-minded members felt the time had come for placing professional scientists rather than aristocratic dilettantes in the chair. So Whewell brings his comparison down a social class, likening his Tables to the various offices within a merchant's accounting office. Here is interconnectedness with specialization and a scientific division of labor, as well as an exemplification of the fact that the truth of the particulars makes certain the truth of the general proposition. Yet this model provides too much equality of genius and equality of task—if Whewell was somewhat uncomfortable with a "scientific nobility," he nevertheless wanted to accord status to those who did more than collect facts and figures. As Whewell continually insisted, induction was something more than the mere sum of its parts. Despite defending this metaphor against those who would call it insufficiently dignified, Whewell could not have liked the implication that science is primarily justified by its practical applications. He therefore shifts to the image of the "Treasury of Science," more dignified than a merchant's accounting office but altogether too suggestive of facts accumulated without reference to the mental element that must organize and coordinate such a Treasury.

Whewell's struggle to find an acceptable replacement for Bacon's model of ascent and descent is also evident in how he deals with the powerful image of Bacon in Cowley's ode. Whewell does not adopt the image without altering it significantly. In his introduction to the *History,* Whewell effaces Bacon: "The eminence on which we stand may enable us to see the land of promise, as well as the wilderness through which we have passed."[108] Whewell's point is clear: the wilderness is the history of the sciences, and the promised land represents the rapid progress of future science. But Whewell's use of this image can also be read as self-referential. Whewell's *History* has become Mount Pisgah, and Whewell is now Moses, inviting the reader to imagine having completed her reading of the *History,* surveying where she has come from (the text of the *History* itself) and where she is going to (the *Philosophy*). Though violence must be done to biblical chronology, Whewell installs himself as the new lawgiver, soon to offer the new dispensation, the renovation of Bacon's *Novum Organum*.

In the *Philosophy,* Whewell again challenges Bacon's status as lawgiver, and again implicitly—given the goal of the work—replaces Bacon with himself. Bacon, says Whewell, "is usually spoken of, at least in this country, as a teacher who not only commenced, but in a great measure completed, the Philosophy of Induction. He is considered not only as having asserted some general principles, but laid down the special rules of scientific investigation; as not only one of the Founders, but the supreme Legislator of the modern Republic of Science; not only the Hercules who slew the monsters that obstructed the earlier traveller, but the Solon who established a constitution fitted for all future time."[109] Bacon is a hero, a slayer of Scholastic monsters, but not a Solon. He requires a Whewell to revise his special rules, for it is only his general principles that continue to have merit.[110]

In Cowley's ode, Bacon is a civil lawgiver, lord chancellor of both king and nature. As Moses he is not only a religious lawgiver as well, but also a liberator and a prophet; having led science from the bondage of Aristotle and the Schoolmen, he shows it the domains it will subdue and occupy. The power of Cowley's image derives from the depth of its associations, but as the example of Whewell demonstrates, *how* this image is appropriated by subsequent writers is revealing about the fortunes of Bacon in the nineteenth century.

Early invocations of the Moses figure tend to isolate and endorse one or two of these associations. Baden Powell's favorable assessment of Bacon's method and influence in 1830 exploits the notion of Bacon as liberator and prophet: "Bacon, from his lofty elevation, took a complete survey of the rich territory of the promised land; but expired, like Moses on Mount Nebo, without himself entering it."[111] For Hallam, Bacon is both liberator and lawgiver, for "he may be compared to those liberators of nations, who have given them laws by which they might govern themselves, and retained no homage but their gratitude."[112] Hallam's imperial liberators sound peculiarly English, carrying their divinely ordained mission into other realms of knowledge much as the Israelites entered the Promised Land or, in Bacon's own time, the European powers launched voyages of discovery and conquest.[113] This imperial theme was congenial to Herschel as well, who likened the extension and progress of scientific knowledge to the conquering and ruling of another land.[114] Somewhat earlier, Playfair had turned these images against the scientists of the Middle Ages, who in his view were in too much of a hurry to become conquerors and lawgivers in their own right—they were like "adventurers in an unexplored country, given up to the guidance of the imagination, . . . in too great haste to become legislators themselves."[115]

As the century progressed, however, the Bacon and Moses references became more and more qualified, testifying, much like the discussions of Bacon's method, to a lingering but increasingly vague sense of Bacon's authority. Macaulay praises Cowley's image, but his endorsement rings rather hollow given the criticisms that elsewhere in the essay strike at Bacon's claims as a methodological liberator or lawgiver. Although he refers to Bacon the lawgiver, Macaulay uses language that celebrates Bacon the prophet:

> Cowley, who was among the most ardent, and not among the least discerning followers of the new philosophy, has, in one of his finest poems, compared Bacon to Moses standing on Mount Pisgah. It is to Bacon, we think, as he appears in the first book of the *Novum Organum*, that the comparison appeals with peculiar felicity. There we see the great Law-giver looking round from his lonely elevation on an infinite expanse;—behind him a wilderness of dreary sands and bitter waters in which successive generations have sojourned, always moving, yet never advancing, reaping no harvest and building no abiding city,—before him a goodly land, a land of promise, a land flowing with milk and honey. While the multitude below saw only the flat sterile desert in which they had so long wandered, bounded on every side by a near horizon, or diversified only by some deceitful mirage, he was gazing from a far higher stand, on a far lovelier country,—following with his eye the long course of fertilizing rivers, through ample pastures, and under the bridges of great capitals. . . .[116]

This pattern occurs frequently. Lewes, for example, affirms Bacon's patriarchal status as the "Father of Experimental Philosophy," but the method of the "great Reformer" is only of historical interest; it "excites no admiration for its present intrinsic value," for "whatever the value of Bacon's works, Bacon's Method is useless."[117] Ellis also refers to the "happy image" of Bacon the Mosaic lawgiver, despite having declared Bacon's method "practically useless." He prefers Bacon's own image of the explorer sailing across an apparently boundless sea, but unlike earlier writers Ellis emphasizes the impossibility of subduing the whole.[118] Fowler and Nichol, supportive of Bacon but aware of the criticism of the details of his method, ignore, like Whewell and Powell, Bacon the lawgiver, focusing instead on Bacon the prophet. Bacon, says Fowler, "stood like a prophet, on the verge of the promised land, bidding men leave, without regret, the desert that was behind them, and enter with joyfulness and hopefulness on the rich inheritance that was spread out before them." Nichol, quoting Cowley, calls Bacon "the prophet of things Newton revealed."[119]

For those who dismissed Bacon or replaced him with Newton, Cowley's image could easily be rejected or transferred. In Liebig's view, Bacon only seems to have brought us to a land of milk and honey; in fact, we are back in the wilderness: "his 'Novum Organum' . . . resembles, in its origin, a merry spring bursting forth out of the ground. It gives us reason to hope that we shall be led by it, through green and flowery meadows, to cool shady woods, to a brook with mills beside it, and, at last, to a stream bearing ships upon its water; but, instead of this, the wanderer is brought into a desert, where there is no life, and where the rivulet vanishes amid the barren sand."[120] Merz, on the other hand, agrees that we "may look upon Lord Bacon as one who inspects a large and newly discovered land, laying plans for the development of its resources and the gathering its riches." But after quoting Cowley's ode and Ellis' assessment that Bacon's merits belong to the spirit rather than the precepts of scientific method, Merz argues that Bacon's achievement is inferior to that of Newton, who "unites" the "first labours of the early pioneers" into "a systematic exploration, and sinks the main shaft which reaches the lode of rich ore," which in turn "opens out the wealth of the mine and marks out the work for his followers." If Bacon is a prophet, he is *only* a prophet. The true lawgiver is Newton, the man who "patiently drew up the first simple rules and gave the first brilliant application. More than the unfinished and wearisome pages of Bacon's 'Novum Organum' does the 'Principia' deserve to be placed . . . as a model work of scientific inquiry."[121]

Although Merz's criticism of the "unfinished and wearisome" *Novum Organum* approaches the extreme, it reveals just how meaningless the praise of Bacon's contributions to the "spirit" of science could be. Herschel and Whewell and Mill had all come to an essentially similar conclusion in the 1830s and 1840s, but Bacon was then still considered worthy of updating, the *Novum Organum* still worthy of study. Those earlier revisions had been a sign of continuing influence and authority, but they nonetheless prepared the way for the faint praise and outright dismissals that followed.

As Bacon's stock, both methodological and general, went down, that of Newton and Descartes went up. In the early part of the century, the only challenge faced by those who asserted that Newton's success was due to Bacon's precepts was to explain Newton's failure to acknowledge Bacon's influence. By the end of the century, however, denials of methodological links between Bacon and Newton had become far more common—Newton had succeeded precisely because he had not tried to follow the sterile mechanism of Baconian induction. De Morgan's editor, commenting in 1915 on De Morgan's emphatic 1872 claim that Newton had not taken Bacon for his methodological master, contended that "De Morgan never wrote a more suggestive sentence. Its message is not for his generation alone."[122] And Descartes, the practitioner of a method labeled "perverse," "diametrically opposite" to Bacon's, and "pure a priori deduction" by Playfair, was hailed a half century later by Huxley as "the father of modern philosophy."[123]

The "manifold debate" which Robertson described as "circl[ing] round the name of Bacon" in the nineteenth century was indeed a *manifold* debate, reaching well beyond a small group of philosophically inclined scientists and scientifically inclined philosophers. One of its core components, arguably its primary core component, was this discussion of Baconian induction, a discussion invariably motivated by an interest in scientific method in general and bound up in contemporary disputes about specific scientific theories and the shape of scientific institutions. The pattern of the methodological discussion we have traced is also evident in these more overtly practical issues. The public controversy over Darwin's theory of natural selection, for example, was focused on method as well as content. As *Blackwood's* used Bacon to discredit him, *Punch* invoked Newton:

> "Hypotheses non fingo,"
> Sir Isaac Newton said
> And that was true by Jingo!

As proof demonstrat*ed*
But Darwin's speculation
Is of another sort
'Tis one which demonstration
In nowise doth support.[124]

Despite Darwin's disingenuous claim that he "worked on true Baconian principles, and without any theory collected facts on a wholesale scale," Huxley, Lewes, and Mill had to defend him from those who, in Huxley's words, "prate learnedly about Mr. Darwin's method, which is not inductive enough, not Baconian enough, forsooth, for them," yet within thirty years Darwin provided Karl Pearson with a typical example of scientific method.[125] Institutionally, the British Association for the Advancement of Science, founded in 1831 with Bacon's writings as its inspiration, was quickly controlled by Whewell and others of the "Cambridge Network," who manipulated its naive Baconian fact collecting to their own theoretical ends.[126] In a presidential address dripping with Whewellian terminology, Prince Albert declared in 1859 that "the Association . . . points out *what* is to be observed, and *how* it is to be observed."[127]

From Prince Albert to the pages of *Punch*, whether Bacon was revised or repudiated, agreement was virtually unanimous that science was far more imaginative than naive Baconianism would allow. The remainder of this study considers the ways in which literary artists became involved, directly or indirectly, in this debate, how they appropriated and responded to specific scientific theories within their own imaginative discourse, and how they conceived of the relationship between scientific and literary methodologies. And it explores the reciprocal appropriation of literature by science—not to prove that science is just another literary construct, but to suggest that science and literature are inextricable elements of a single culture and share, more often than we have led ourselves to expect, a common discourse.

# 2

# Reweaving the Rainbow

## Romantic Methodologies of Poetry and Science

In a 1970 article for *Nature* about the friendship between Wordsworth and the Irish mathematician W. R. Hamilton, George Dodd begins with the comment that "[i]t comes as a surprise to most scientists to think of Wordsworth—the leader of the romantic poets—as having an interest in and expressing opinions on scientific matters."[1] What makes such a statement truly surprising in the pages of *Nature*, however, is the fact that *Nature*'s motto—appearing with some prominence at least through 1963 and discussed in detail in the journal's centenary issue the year prior to Dodd's article—is taken from Wordsworth himself: "To the solid ground / Of nature trusts the Mind that builds for aye."[2]

Dodd's own apparent unawareness of the historical connection between Wordsworth and *Nature* ironically underscores his claim that for "most scientists" science and poetry are fundamentally at odds. Expressed schematically, this antithesis suggests that poetry (indeed, all artistic creation) is personal, subjective, spiritual, emotional, and organic, while science is impersonal, objective, physical, logical, and mechanistic. But such assumptions about the hostility between science and Romantic poetry have not been the exclusive property of scientists; literary critics, in spite of a long tradition in Romantic studies exploring the place of science in Romantic thought and the involvement of individual poets with science, have helped to create and institutionalize them.[3] In fact, had Dodd turned to Charles Pittman's study of Wordsworth and Hamilton, he would have discovered that Wordsworth's relationship with Hamilton, by increasing his interest in science and leading him to incorporate scientific knowledge into his poetry, was supposedly a direct cause of Wordsworth's poetic decline.[4]

Pittman is, admittedly, of another era. Subsequent studies of the Romantics have made clear that the science/poetry antithesis was applied not to science in general, but to Newtonian science in particular, not to scientific method in general, but to naive Baconianism in particular. For Keats, Newtonian science was "cold philosophy," and it was epitomized by Newton's theories of color and of light, which Keats felt served only to "unweave the rainbow," to reduce the mysterious charms of nature into a "dull catalogue of common things."[5] Keats's comment strikes at the dual heart of the Romantic objections to Newtonian science and Baconian method: corpuscular theories of matter turned nature into a giant mechanism with no place for the individual human will and human imagination, while collecting and classifying for the sake of collecting and classifying ignored the relationship between natural beings, especially the human organism, and their environments. The Romantics wanted both science and poetry to explore an organic universe rather than a machine, a universe where reality is created—or at least half created—by the observing mind. And this element of replacement is important: Goethe's *Die Farbenlehre* (Theory of Colors) not only challenged Newton's account of the rainbow, but offered an alternative explanation. As Hegel bluntly declared in his defense of Goethe's theory, "[n]o artist is stooge enough to be a Newtonian."[6]

Yet even modern critics well aware of this Romantic critique of the Baconian/Newtonian tradition tend to fall back on more subtle forms of the science/poetry dichotomy. In a recent discussion of how to counter "the artificial opposition between arts and sciences that has become so deeply entrenched in modern culture," Joel Black argues that we must "trace this opposition back to its roots in scientific discourse itself" and "reveal how the new rationalist discourse of science required and created the cultural fiction of just such an antithesis between artistic subjectivity and scientific objectivity in order to validate its own absolute claim to the truth."[7] Although he wants to put an end to the "artificial opposition" of science and poetry, and although he sees that opposition lurking even in the nooks and crannies of Hans Eichner's argument about the Romantic reaction to Newtonian science, Black sees *science* as the villain. The source of what Black calls "our modern-day cultural schizophrenia" lies in "scientific discourse," which created this cultural antithesis for its own sinister and hegemonic ends. Romantic literary discourse, which by implication had no role in the creation of this schizophrenia, is on the other hand cast as the pioneering and unappreciated doctor endeavoring "to diagnose and treat this condition."

Black's language shares, as we shall see, a common feature of Romantic literary discourse: the invocation of scientific language and metaphor to describe the poetic process or to expound the difference between science and poetry. But more important, Black's language, as it reveals just how difficult it can be to avoid the science/poetry antithesis, suggests a somewhat different approach. Rather than assuming that this antithesis has its origins exclusively in scientific discourse, I want to consider the possibility that the Romantics themselves had a hand in shaping and appropriating it. By the same token, rather than assuming that scientists simply and uniformly endorsed a standard of absolute truth that excluded poetry, I want to suggest that many important scientists throughout the nineteenth century did not see the two enterprises as antithetical. This is not to deny that scientific discourse did indeed help to create and to exploit hostility between poetry and science—it is merely to contend that the issues are more complicated and more laden with tension than is usually admitted, perhaps especially in the Romantic era.

In this chapter I explore these tensions in the works of two of the period's best-known literary figures and its leading scientist—Wordsworth, Coleridge, and Humphry Davy—each of whom devoted considerable attention to methodological questions about what it meant for a scientist to do science and what it meant for a poet to write poetry. The connections among this trio, and the interest that they took in each other's work, are well documented (the close friendship and subsequent falling out between Davy and Coleridge, Davy's seeing the *Lyrical Ballads* through the press, Coleridge's aborted plan for a chemical laboratory in the Lakes, Davy's recruitment of Coleridge for a series of lectures on poetry at the Royal Institution), yet their responses to the questions they posed about the relationship of scientific and literary method reveal important differences in emphasis as well as broad agreement. By locating these men in the context of the early stirrings of the nineteenth century's debate about the Baconian/Newtonian tradition, I hope to show that Romantic poetics and scientific methodology are in many ways not at odds but rather a part of a unified cultural movement, that they borrow from each other's discourses as they tentatively approach a similar position, the tensions of which will continue to plague the Victorians and will make possible the assertions of cultural schizophrenia that continue to plague us.

## WORDSWORTH: TRANSFIGURING SCIENCE

In what is probably the most famous discussion of science and poetry in English literature, Wordsworth compares the endeavors of the poet to those of the "man of science" in the preface to the 1802 edition of the *Lyrical Ballads*. Arguing that the poet comes to a knowledge of nature more rapidly and instinctively than the man of science, Wordsworth claims that poetry encompasses science, that poetry is "the first and last of all knowledge."[8] But, he continues,

> If the labours of men of science should ever create any material revolution, direct or indirect, in our condition, and in the impressions which we habitually receive, the poet will sleep then no more than at present, but he will be ready to follow the steps of the man of science, not only in the general indirect effects, but he will be at his side, carrying sensation into the midst of the objects of science itself. The remotest discoveries of the chemist, the botanist, or mineralogist, will be as proper objects of the poet's art as any upon which it can be employed, if the time should ever come when these things shall be familiar to us, and the relations under which they are contemplated by the followers of these respected sciences shall be manifestly and palpably material to us as enjoying and suffering beings. If the time should ever come when what is now called science, thus familiarized to men, shall be ready to put on, as it were, a form of flesh and blood, the poet will lend his divine spirit to aid the transfiguration, and will welcome the being thus produced, as a dear and genuine inmate of the household of men. (1:141)

This passage has often been cited, and continues to be cited, as an eloquent expression of the compatibility of science and poetry. Yet the reiterated condition that the full expression of this compatibility will be possible only when "the objects of science" have become "manifestly and palpably material to us" indicates that for Wordsworth such is not the case now. The mere separation of "poet" and "man of science" is itself significant as an indication of some sort of difference, but by emphasizing as hypothetical that postrevolutionary day when the poet follows the man of science, Wordsworth implies that the man of science is currently following the poet. The "transfiguration" of science will thus require the "divine spirit" of the poet, who will then "welcome" this new being into "the household of men." This metaphor accords science, in a backhanded way, its own divine status, but what is missing from science is that which poetry already has: a human form "of flesh and blood." Any transformation of science into the first and last of all knowledge will, in other words, be not so much the scientizing of poetry as the poetizing of science.

But a further clue that accounts for some of these gaps between poet and man of science occurs in Wordsworth's conception of "what is now called science." In a footnote which rejects the "contradistinction" of poetry and prose, Wordsworth says that "the true antithesis" is "the more philosophical one of poetry and matter of fact, or science" (1:135). Wordsworth added the phrase "matter of fact" to the expanded version of the preface, a conscious elaboration that gives us a sense of what he means here by "science." If "science" equals "matter of fact," then "science" closely corresponds to what we have seen identified as "the common notion" of Baconianism: the collection of empirical data without theoretical speculation, a monotonous and purely logical exercise that attempts to find the true by eliminating the false. Botany, mineralogy, and even chemistry were primarily classificatory sciences; although the work of a Davy suggested that chemistry had the potential to engage us as "enjoying and suffering beings," the bulk of chemists was content with the determination of boiling points and reaction ratios. If this is science, then it is easy to see why poetry can be viewed as its opposite.[9]

M. H. Abrams points out that Wordsworth's contrast of poetry to "matter of fact, or science," was one of the most widely adopted propositions of the preface.[10] It is difficult to say, however, just how Wordsworth wanted this contrast to be taken. While it now seems clear that Wordsworth was here associating science with the worst excesses of Baconianism, the passage does not spell this out. Coupled with the ambiguities about the transfiguration of science, the contrast leaves open the possibility that Wordsworth is not qualifying what he means by science. Since the preface is highly rhetorical, and since Wordsworth in this section is most interested in establishing the nature of *poetry*, he very possibly welcomed this broader interpretation as a way of making the contrast more forceful and more advantageous to poetry. Yet it is precisely this broader interpretation of science that later writers found most congenial.

In his 1844 essay "What Is Poetry?" Leigh Hunt declares that "[p]oetry begins where matter of fact or of science ceases to be merely such, and to exhibit a further truth; that is to say, the connexion it has with the world of emotion, and its power to produce imaginative pleasure."[11] Like Wordsworth but even more bluntly, Hunt argues that poetry goes beyond "matter of fact or of science." Poetry is concerned with truth and knowledge, but it offers an additional truth in its connection to the emotions and its ability to produce pleasure; the emotionless scientist stops at the reporting of a fact, but the poet records his response to it. To the degree that that emotional response is communicated genuinely and sincerely, imaginative pleasure (which Hunt does not seem to associate with science at all) and additional truth are generated.

To elucidate his position, Hunt offers an example which, not surprisingly, invokes the classificatory endeavors of botany: "Inquiring of a gardener, for instance, what flower we see yonder, he answers, 'A lily.' This is matter of fact. The botanist pronounces it to be of the order of 'Hexandria monogynia.' This is matter of science. It is the 'lady' of the garden, says Spenser; and here we begin to have a poetical sense of its fairness and grace. It is 'The plant and flower of light,' says Ben Jonson; and poetry then shows us the beauty of the flower in all its mystery and splendour." Indeed, Hunt contends that poetic intuition often anticipates matters of fact only later demonstrated by science, as that light, like the lily, is white.[12] Although he distinguishes between the gardener's matter of fact and the botanist's matter of science, Hunt makes the latter even more distanced from the responses of Spenser and Jonson by its specialized, classificatory nomenclature. The sense in Wordsworth that the emotionless Baconian scientist is a bad scientist has been lost in Hunt's rhetoric, where the botanist stands for all scientists, and the botanist's language of classification, shorn even of the descriptive element on which the final classification must depend, is made to stand against that of Spenser and Jonson.

An even more striking example is De Quincey's distinction between the literature of knowledge and the literature of power, a distinction he ascribed to "many years' conversation with Mr. Wordsworth."[13] First articulated in his "Letters to a Young Man" (1823) and later expanded in his essay "The Poetry of Pope" (1848), De Quincey's distinction is much more absolute than Wordsworth's, however. In the preface to the *Lyrical Ballads*, Wordsworth argues that the poet's only restriction is the necessity of producing immediate pleasure, but that knowledge itself is impossible without pleasure: "We have no knowledge . . . but what has been built up by pleasure, and exists in us by pleasure alone" (1:140). Whether for poet or man of science, "knowledge . . . is pleasure"; even the anatomist, struggling with the "disgusts" of his occupation, feels that "where he has no pleasure he has no knowledge." The difference, in a reversal of our common modern conception, is that while the poet is a communal figure whose knowledge comes from within, the man of science is a solitary figure whose knowledge comes from without. The poet's knowledge "cleaves to us as a necessary part of our existence, our natural and unalienable inheritance," while the scientist's knowledge is "a personal and individual acquisition," developed "through labor and length of time . . . by conversing with those particular parts of nature which are the objects of his studies." Like Wordsworth, De Quincey does not try to distinguish poetry from other forms of discourse according to an antithesis of pleasure and knowledge, but in his claim that he is following Wordsworth by invoking the "true antithesis" (10:48) of knowledge and power, De Quincey transforms the differences between poetry and science into a more absolute schism. While acknowledging in "The Poetry of Pope" that knowledge and power often blend in an individual work, De Quincey believes that they are not only "capable . . . of a severe insulation," but "naturally fitted for reciprocal repulsion" (11:54). The function of the literature of knowledge is to teach, but the function of the literature of power is to move. The truths involved in conveying information are subordinate to the truths associated with human emotions, human moral capacities, and human pleasure. And De Quincey's examples make clear that science does not share these more important truths.

"What do you learn from 'Paradise Lost'?" asks De Quincey. "Nothing at all. What do you learn from a cookery-book? Something new, something that you did not know before, in every paragraph. But would you therefore put the wretched cookery-book on a higher level of estimation than the divine poem?" (11:55–56). The absurdity of comparing a cookbook with *Paradise Lost* goes beyond even Hunt's similar strategy, and thus De Quincey must adopt a different argument when he replaces "the wretched cookery-book" with Newton's *Principia*. Although the *Principia* is an example of "[t]he very highest work that has ever existed in the Literature of Knowledge," it is "but a *provisional* work" (11:57). The information conveyed by the literature of knowledge may be important, but it is, ultimately, information nonetheless, not associated with the truths of a higher and permanent sphere, and therefore susceptible to revision and reinterpretation, rejection and replacement: "Let its teaching be even partially revised, let it be but expanded,—nay, even let its teaching be placed in a better order,—and instantly it is superseded." In the case of the *Principia*, De Quincey argues that even though its truths continue to provide the basis for contemporary mechanics, such expansion has made it obsolete. Turning the supposed progression of scientific knowledge against Newton, De Quincey cites the work of Laplace—generally used to celebrate Newton—as eclipsing both the *Principia* and Newton himself: "as soon as a La Place . . . builds higher upon the foundations laid by this book, effectually he throws it out of the sunshine into decay and darkness; by weapons won from this book he superannuates and destroys this book, so that the name of Newton remains as a mere *nominis umbra*, but his book, as a living power, has transmigrated into other forms." As with Wordsworth's transfiguration of science, De Quincey's transmigration of the scientific soul is a backhanded compliment; transmigration equals superannuation and destruction, and superannuation and destruction are not the fate of great literary works.

If the *Principia* can be located in the same category with cookbooks, then we have a severe disjunction indeed between poetry and science, and a disjunction claimed to derive from Wordsworth. Yet in his discussion of the stone-and-shell allegory in *The Prelude*, De Quincey argues that the passage shows Wordsworth's veneration for poetry and mathematics as the "two hemispheres" of "the total world of human power" (2:268). This is especially surprising given Wordsworth's specific identification of the stone with Euclid's *Elements*, which, given its status as a textbook, would seem a perfect example for De Quincey of the literature of knowledge rather than the literature of power. Moreover, since De Quincey saw the manuscript of Wordsworth's essay "The Convention of Cintra" (1809) through the press, he would have seen the following passage (subsequently canceled): "Lord Bacon two hundred years ago announced that knowledge was power and strenuously recommended the process of experiment and induction for the attainment of knowledge. But the mind of this Philosopher was comprehensive and sublime and must have had intimate communion of the truth of which the experimentalists who deem themselves his disciples are for the most part ignorant viz. that knowledge of facts conferring power over the combinations of things in the material world has no determinate connection with power over the faculties of the mind" (1:324). Wordsworth, like Coleridge, argues that Bacon knew what he was about even if those "experimentalists who deem themselves his disciples" do not. Knowledge and power are one, but knowledge of the "facts" of the material world should not be thought to render the mind unimportant. Knowledge of things does not bring power without the active participation of the mental faculties; thus Wordsworth goes on to lament that "all those products of knowledge which are confined to gross—definite—and tangible objects, have, with the aid of Experimental Philosophy, been every day putting on more brilliant colours; the splendour of the Imagination has been fading" (1:325). And overattention to material things at the expense of mind and imagination eventually leads to moral decline.

Yet where Wordsworth's language suggests the desire to restore some sort of balance, De Quincey creates a hierarchy by separating knowledge and power. Anything to do with facts and things, whether a cookbook or the *Principia*, is mere knowledge, mere information to be taught. Only the literature of power is concerned with "the moral concerns of man, with ideals, and with imagination" (11:56–57). Thus for De Quincey the "speaking symbol" of the literature of knowledge is the encyclopedia, which "before one generation has passed . . . is superannuated" and which "speaks through the dead memory and unimpassioned understanding . . ." (11:59). Again, not surprisingly, De Quincey exemplifies the literature of knowledge with a work easily caricatured for its lack of imagination, its obsession with conveying information that is nonetheless rapidly out of date, its reliance on a system of classification (the alphabetical) that is not really a system at all. Yet, later in the essay, when he is concerned to demonstrate that didactic poetry, in particular Pope's "Essay on Man," is a contradiction in terms, De Quincey relies on classificatory language inevitably suggestive of the Linnaean botany that was so abhorrent to him and to other Romantics: "No poetry can have the function of teaching. It is impossible that a variety of species should contradict the very purpose which contradistinguishes its genus" (11:88).

Wordsworth's antithesis between poetry and "matter of fact, or science," could thus be easily simplified into an absolute hierarchical disjunction by followers claiming to derive their position from Wordsworth. Yet if Wordsworth's language in his discussion of the relationship of poetry and science is, as I have suggested, more complex and contradictory, it is worth asking what the source of those complexities and contradictions may be. M. H. Abrams argues that Wordsworth was more immersed in eighteenth-century thought than most of his contemporaries, and that we must look to his poetry rather than his criticism for the transition away from the eighteenth-century view of man and nature to the position that the mind is active and creative in perception.[14] Yet we have already begun to see even in Wordsworth's criticism both a willingness to employ and a desire to critique the eighteenth century's formulation of the Baconian/Newtonian tradition. Praising Bacon, he excoriates Bacon's followers for making the mind of small importance compared with things. Maintaining an emphasis on poetry's need to produce pleasure, he contends that science, too, is a source of pleasure, but that poetry, not science, is the first and last of all knowledge. Arguing that poetry is more properly contrasted to science, he nonetheless considers science as it should be practiced to be divine, and denies that the two are mutually exclusive. And such conflicts are evident in Wordsworth's prefaces even where he is talking not specifically about the relationship of science and poetry, but about the principles of his own poetry.

## THE PREFACE TO THE *LYRICAL BALLADS*: THE SCIENCE OF POETIC INDUCTION

In the second sentence of both the 1800 and the 1802 versions of the preface, Wordsworth says that the *Lyrical Ballads* are "an experiment, which, I hoped, might be of some use to ascertain, how far by fitting to metrical arrangement a selection of the real language of men in a state of vivid sensation, that sort of pleasure and that quantity of pleasure may be imparted, which a poet may rationally endeavour to impart" (1:118–19). The word "experiment" is the first word used to describe both the *Lyrical Ballads* as a whole and the individual poems contained within it. (It is also used by Wordsworth in the brief "Advertisement" attached to the first edition in 1798.) The poems that comprise this volume test a new hypothesis about the nature and subject matter of poetry, with an eye toward "ascertaining" how successful this new approach might be, in terms of both execution and reception. Presumably, the degree of success would be used as a guide in determining whether this hypothesis should be embraced, discarded, or modified by its poetic practitioners. In fact, this conception of the *Lyrical Ballads* had already affected the exclusion and inclusion of particular poems, such as Coleridge's "Christabel."[15] Furthermore, this group of experiments will provide, according to Wordsworth's statement, results which can be measured in both qualitative and quantitative terms against a reasonable if vaguely defined standard: the quality and quantity of pleasure "which a poet may rationally endeavour to impart."

Although these poetic experiments begin not with the aimless collection of data but with the testing of a hypothesis, and despite Wordsworth's criticism of naive Baconianism, there is a sort of Baconian insistence that the subject matter of the *Lyrical Ballads* is grounded in observation and that truths of the *Lyrical Ballads* are arrived at inductively and objectively. Wordsworth stresses that his "principal object" is the description of "incidents and situations from common life" (1:123). In language weighted with the terminology of chemistry and physics, he declares that the poet "considers man and the objects that surround him as acting and re-acting upon each other" (1:140). The poetic vocabulary most appropriate for depicting these incidents of common life is, of course, the oft-repeated "selection of language really used by men" (1:123). Although complete access to real language and real emotion is impossible, the poet does have a greater claim on objective truth because he can reproduce this language and emotion so vividly through contemplation. Poetry "takes its origin from emotion recollected in tranquility" (1:149), in the ability of the poet to distance himself, like the scientist, from that which he observes, to prevent the volubility of personality and emotions from obscuring the "true" experience. What differentiates the poet from the scientist (and the distinction is again telling as an indication of Wordsworth's notion of science) is that the poet does not try to exorcize the emotions permanently, but instead endeavors to recall them in a controlled fashion—the poet's subject matter is, ultimately, the mind and emotions of the poet. But Wordsworth is able to describe this process only by another metaphor derived from chemistry: "the emotion is contemplated till by a species of re-action the tranquility gradually disappears, and an emotion, kindred to that which was before the subject of contemplation, is gradually formed" (1:149).

This claim of privileged access to a world that is objectively present is reflected in the poems' "language of real men." If the events and language of common life are not the sort usually encountered by the well-read, well-bred audience of poetry, they have the advantage of being more meaningful, from a "scientific" perspective, precisely because they are more typical of the human lot. The choice of "humble and rustic life" makes it possible to study "the essential passions of the heart" (1:125) in their simplest, purest, most durable form. Because people living such a life are unreflective and relatively free of contact with a more complicated, diverse, and "unnatural" social order, they themselves provide unbiased commentary, in simple but powerful language, on these "essential passions."

Through such extended observation of the common and simple, poetry elucidates truths which are universal and sophisticated, "the primary laws of our nature" (1:123). Like good inductive science, it begins in individual events, with individual people, but "its object is truth, not individual and local, but general and operative" (1:139). Indeed, Wordsworth defines knowledge in the preface as "general principles drawn from the contemplation of particular facts," and although this process is always "built up" by pleasure, both poetic and scientific knowledge are derived inductively through pleasure: "The man of science, the chemist and mathematician, whatever difficulties and disgusts they may have had to struggle with, know and feel this"

(1:140). In offering his poetic theory, Wordsworth is careful to present both these primary laws of human nature and his own beliefs about poetry as emerging from his observations and his poetic experiments. The *Lyrical Ballads* are not made to conform to a group of a priori assertions as eighteenth-century poetry is. Borrowing from the authority of scientific laws inductively established, he knows that his own poetry—so different from what has come before it—will be more apt to be accepted if it is seen as purer and more "real" rather than simply idiosyncratic. Even where his poems fail, Wordsworth says that the *Lyrical Ballads* are distinguished by a "worthy purpose" (1:125) not imposed on the poems in advance but developed during the process of observation, meditation, and composition.

## THE 1815 PREFACE: THE SCIENCE OF POETIC CLASSIFICATION

Like the preface to the *Lyrical Ballads*, Wordsworth's preface to the 1815 edition of his *Poems* is characterized by tensions between an "inductivist" poetic philosophy and a desire to avoid the narrow, "matter of fact" basis of science, in spite of its very different approach. Wordsworth continues to cast his poetic methodology in scientific form, but on this occasion he develops an elaborate classificatory system based on the various steps of the poetic process and the different types of poetry, and then locates his own work within that scheme. Given Romantic hostility toward the classificatory sciences, Wordsworth's endeavor may seem odd, for he shared Keats's contempt for science's "dull catalogue of common things" and himself lampooned, in two oft-quoted passages, the natural philosopher who would "peep and botanize / Upon his mother's grave" ("A Poet's Epitaph" [1800]) and the geologist who "with pocket-hammer smites the edge / Of luckless rock or prominent stone, . . . / The substance classes by some barbarous name, / And hurries on . . . " (*The Excursion* 3:178–85). Yet Wordsworth had declared in his note to "The Thorn" (1800) that poetry is the "science of feelings," and as with any science, poetry needs its own system of classification, though one that avoids being an inert and "barbarous" taxonomic scheme.[16]

To do so, Wordsworth appropriates the "common view" of the practice of the Baconian scientist and then transcends it. As in the earlier preface, he believes not only that poetry can reveal the laws of human nature, but also that the production of poetry is itself governed by definite and discoverable laws. There are "laws . . . of every species of composition" (3:27), and "the operations of the mind upon those objects and processes of creation or of composition" are themselves "governed by certain fixed laws" (3:31). These laws are best discovered inductively, by the actual production of poetry rather than by the a priori construction of them. And of the six "powers" necessary for "the production of poetry," the first is Observation and Description. Wordsworth calls this dual power "indispensable to a poet"—there can be no poetry without "the ability to observe with accuracy things as they are in themselves, and with fidelity to describe them, unmodified by any passion or feeling existing in the mind of the describer" (3:26). The poet, like the Baconian scientist, endeavors first to render accurately an external reality which can be seen as it really is by the unbiased observer.

But Wordsworth goes on, even more forcefully than in the preface to the *Lyrical Ballads*, to emphasize that observation and description merely provide the raw material transformed by the other powers into poetry. Although they are essential for initiating the poetic process, they are only a beginning. Fidelity of description—pure mimesis—does not a poem make. "This power," says Wordsworth, "is one which [the poet] employs only in submission to necessity, and never for a continuance of time: as its exercise supposes all the higher qualities of the mind be passive, and in a state of subjection to external objects" (3:26). By using the power which Wordsworth calls Sensibility, the poet's mind ceases its passive observation and becomes active, contemplating objects both "as they exist in themselves" and "as re-acted upon" by the poetic mind itself (3:26). Reflection enables the poet to dwell on the values of actions and things, thoughts and emotions, to make connections among them. Imagination and Fancy, and Invention, then shape the material supplied by Observation, modifying and creating. The final power, Judgment, determines when and how the other powers should be employed. Looking back at *Nature*'s Wordsworthian motto, we can see that it implicitly acknowledges both the necessity and the ultimate insufficiency of passive observation and description of the natural world: nature is "the solid ground," but it is the ground for "the *Mind* that builds for aye"—and the aye/eye pun stresses that observation can have no permanence without the active labors of the mind. Such a view is in keeping with the content of the sonnet itself, which criticizes that "VOLANT Tribe of Bards" who do not base their poetry in "the solid ground of nature"; their "nests of clay," hung where "the flattering Zephyrs round them play," are "quickly from that aery hold unbound, / Dust for oblivion!"

It is important to recognize, however, that if Wordsworth's rejection of naive realism and naive Baconianism are related, scientists of the period were also making such rejections in favor of a conception of science remarkably similar to Wordsworth's conception of poetry. In his "Introductory Lecture to the Chemistry of Nature" (1807), Wordsworth's friend Humphry Davy tells his audience that "[t]he *body* of natural science consists of *facts*," that it is "the history of things, and their faithful representation."[17] In the "Introductory Discourse" to his first series of lectures at the Royal Institution in January 1802, Davy asserts that modern chemistry "has shown the external world in its distinct forms" (2:321). As with Wordsworth, Davy assumes that nature is, at a fundamental level, objectively real and therefore that it can be accurately described. But Davy also shares Wordsworth's conception that this is not the end of the matter for science any more than it is for poetry: "[Science] has bestowed upon [man] powers which may be almost called creative; which have enabled him to modify and change the beings surrounding him, and by his experiments to interrogate nature with power, not simply . . . passive and seeking only to understand her operations, but rather . . . active with his own instruments" (2:319). Roger Sharrock has argued that the two editions of the preface to the *Lyrical Ballads* and Davy's "Introductory Discourse" comprise a dialogue about the nature and value of poetry and science.[18] It seems to me that this dialogue, with Coleridge intimately involved, is still going on from Wordsworth's side through the 1815 preface, and that what this dialogue is moving toward is uneasy accommodation. Wordsworth gives poetry a scientific basis in fact, in careful observation and exact description, which is then imaginatively transformed in an inductive manner, from particular instances to general truths; Davy contends that science utilizes this same poetic transformation of empirical data into the elegant laws of nature which in turn reinforce the beauty and harmony of the natural world.

But Wordsworth's uneasiness is also evident when, after distinguishing the six poetic powers and the combinations of them that produce different types of poetry, he discusses the arrangement and classification of his own poems. The arrangement of the poems in the 1815 volume follows two major and separately conceived classificatory schemes: one is based on Wordsworth's psychology of literary creation according to the predominating power in each poem or to the type of poem produced, while the other is a simple chronology modeled on a human life from birth to death.[19] Most of the poems in the volume are located under headings appropriate to one of the schemes: "Poems of the Fancy," "Poems of the Imagination," "Narrative," etc., correspond either to the powers or to the types of the psychological scheme; "Juvenile Pieces," "Poems Referring to the Period of Childhood," "Epitaphs and Elegies," etc., correspond to the human chronology. But some headings—for example, "Poems on the Naming of Places"—bear no clear relation to either system. The resulting arrangement therefore does not seem like much of an arrangement at all. At best, it looks like two unrelated systems thrown together, an effort to impose organization on a group of poems that resists such easy categorization. In W. J. B. Owen's words, "[n]o single basis . . . underlies the classification, and no meaningful interpretation can be given it as a whole."[20] Yet the very lack of a unified and coherent system of classification is itself revealing if we locate Wordsworth's taxonomic efforts within the movement from natural history to biology, from the classical to the modern episteme, traced by Michel Foucault.

Foucault argues that in the seventeenth and eighteenth centuries, knowledge, especially in the sciences, is represented by the table, a grid where each rectangle contains its appropriate member, separated from other members but adjacent to those with which it has similarities.[21] The pieces of the grid are "unencumbered spaces in which things are juxtaposed, . . . non-temporal rectangle[s] in which, stripped of all commentary, of all enveloping language, creatures present themselves one beside another, their surfaces visible, grouped according to their common features, and thus already virtually analyzed, and bearers of nothing but their own individual names."[22] In botany or zoology, for example, the allocation of a particular plant or animal to a particular location in the table is based on physical structure, on the composition and arrangement of parts visible to the classifying eye; the process of classifying is therefore simultaneously the process of naming, of fixing the plant or animal both in the static order of the table and in the discourse associated with it.

In the late eighteenth century, however, the basis of classification in these sciences changes from visible physical structure to organic structure: lungs are important not for their *position* in the body, but for their part in a system of relationships that, taken together, performs the *function* of respiration. These various internal systems—respiratory, circulatory, reproductive, digestive, nervous, skeletal, etc.—are in turn interrelated to form an organic whole which cannot be described or determined by abstracting one organ or one system from the rest. The classifying eye now relates the visible and the invisible: structural prominence no longer necessarily implies functional prominence. The organizing principle of the table shifts from identity and difference to analogy and succession: adjacent members of the table are adjacent not because they share a few physical features, but because of relationships among the members and functions of their organic structures, relationships which indicate not permanence but rather a similar development in time. To classify and to name, once the same process, become separate mental activities, and the visible structure that makes possible the identification of an organism's name in the table now relates to the classificatory system that generated the table only in so far as that visible structure is connected to invisible function.

The Romantics of course disliked the classificatory schemes of seventeenth- and eighteenth-century natural history for the passivity of mind that it implied: classification was a sterile, unimaginative exercise in pigeonholing that took no interest in the relationship between observer and observed or between the object and its environment. This is Wordsworth's classifying geologist, whose passive mind merely observes, describes, and names. The Romantics therefore welcomed what Foucault calls the replacement of classification by anatomy, for anatomy placed greater emphasis on the function of the living organism growing and developing over time. But the Romantics were also aware that the study of internal organs and functional systems required that organisms be dissected. And as Wordsworth put it in "The Tables Turned" (1798), "Our meddling intellect / Mis-shapes the beauteous forms of things:— / We murder to dissect." We do not find the "true" organism by exposing the internal organs.

Yet we recall that as early as the preface to the *Lyrical Ballads*, Wordsworth had made clear that dissection need not be murder: "[h]owever painful may be the objects with which the anatomist's knowledge is connected, he feels that this knowledge is pleasure; and where he has no pleasure he has no knowledge" (1:140). If there is to be pleasure and knowledge, mere description of the visible must be avoided in both poetry and science, and dissection, whether of poetry or of dead "objects," must always maintain a connection to the living organism and to the dissector. Thus in the supplementary essay to the 1815 preface, Wordsworth, likening "genuine poetry" to "pure science," implies that the phenomenology of the poet should also be the phenomenology of the scientist, that both have a duty "to treat things not as they *are*, but as they *appear*; not as they exist in themselves, but as they *seem* to exist to the *senses*, and to the *passions*" (3:63, original emphasis).[23] That dissection is not necessarily a fetish for "things as they are," so long as "the sense . . . be cultivated through the mind," is confirmed by a later (1843) comment in which Wordsworth criticizes those who believe that "the habit of analyzing, decomposing, and anatomizing, is inevitably unfavorable to the perception of beauty," when in fact "the beauty in form of a plant or an animal is not made less but more apparent as a whole by a more accurate insight into its constituent properties and powers."[24]

In the dissection of his poetry Wordsworth is therefore far less concerned with the types of poetry than with the powers which infuse them, with the relation between the poem and "The Growth of a Poet's Mind"; to the "anatomist" of poetry, what matters is not simply outward form. Wordsworth insists that a power is not present in a poem in pure form—the poems are arranged according to which power *predominates* in them—and therefore by implication a poem cannot be chopped up into its respective powers, but must be considered as a unified whole. To maintain a link with the senses and the passions, and to keep the whole before us, Wordsworth yokes to each of his two systems an organic metaphor—the progress of a human life in the one case, the process of a poet's mind in the other. He even extends these schemes to his own career, acknowledging that the arrangement is that not simply of a self-contained work, but of a progressive movement toward his planned great work, *The Recluse*. In this feature, the 1815 preface closely follows the preface to *The Excursion*, in which Wordsworth sketches the general plan of *The Recluse*, the position of *The Excursion* within that plan, and the relation of both to the rest of his work.

In a conversation some years later with W. R. Hamilton about the remarks on science in book 4 of *The Excursion*, Wordsworth defended his distinction between science "in the proper sense of the word" and science "which was a bare collection of facts for their own sake, . . . which waged war with and wished to extinguish Imagination in the mind of man."[25] He distrusted the tradition of Baconian/Newtonian science in so far as its methodology turned both nature and the observer of nature into a sort of machine: nature was not an object to be taken apart and analyzed by a detached observer who, so long as he followed the proper steps, would eventually be left with a natural law in his scientific lap. Nor was it to be taken apart and analyzed for the sole purpose of taking apart and analyzing, of classifying its pieces into the rigid order of a table. Thus when he appropriates scientific discourse for his poetry and poetics, Wordsworth tries to avoid or redefine those aspects of the Baconian/Newtonian tradition which he disliked. His classificatory system in the 1815 preface may not be unified and coherent, but it seeks to be more than—just as "genuine poetry" and "pure science" must be more than—a "bare collection of facts." And as he notes in the preface to the *Lyrical Ballads*, even the poet's enterprise is susceptible to such degeneration into passivity and sterility, especially given the fact that when "compared with the freedom and power of real and substantial action and suffering," the poet's situation is "altogether slavish and mechanical" (1:138). Far from marking him as an opponent of science or a creature of eighteenth-century scientific thought, Wordsworth's revisions, particularly in the 1815 preface, place him within the broader cultural critique of scientific method and classification. Wordsworth wanted to see (and did see, in figures like Davy and Hamilton) a transformation in science and scientific methodology, just as he wanted to see (and did see, in himself and Coleridge and others) a transformation in poetry and poetic methodology, but he was not above exploiting these older views, in either discourse, for his own purposes. If there are tensions, uncertainties, even contradictions in his views, they are characteristic not only of Wordsworth, but of other poets and scientists of the Romantic period, including Coleridge and Davy as well.

## COLERIDGE AND THE SCIENCE OF METHOD

Coleridge's interest in methodology was both more general and more specific than Wordsworth's. His writings, both public and private, are full of such methodological discussions, the most sustained being the "Essay on Method" in *The Friend*. In developing his own grand system of knowledge, Coleridge wanted to link the methods of the fine arts and the sciences with those of philosophy, morals, and religion, and the "Essay on Method" was crucial to that goal. "Our discussion," he wrote, "is confined to Method as employed in the formation of the understanding, and in the constructions of science and literature," but it is also "the basis of all my future philosophical and theological writings."[26] Confirming that assessment in private, Coleridge remarked on at least one occasion that he considered the section of *The Friend* which includes the "Essay on Method" to outweigh all his other works, verse or prose.[27] My own discussion of Coleridge will focus on *The Friend* in part because of this central place that Coleridge himself gave to it, but also because it is here that Coleridge explores most thoroughly the relationship of science and poetry and expounds most systematically his version of Baconian methodology.

As with Wordsworth and the *Lyrical Ballads*, Coleridge stresses that *The Friend* is "an experiment" (1:15). In this preliminary attempt to formulate his all-encompassing philosophy, and especially to lay the groundwork of its methodological foundations, Coleridge distrusts ungrounded speculation but recognizes the limits of observation. If he were to choose his own judge, Coleridge says, he would want a man "whose knowledge and opinions had for the greater part been acquired experimentally; and the practical habits of whose life had put him on guard with respect to all speculative reasoning, without rendering him insensible to the desirableness of principles more secure than the shifting rules and theories generalized from observations merely empirical . . ." (1:4). As so often in Coleridge's thought, principles are to be grounded inductively in experience and experiment, but "the development of general truth for a general purpose" (1:389) is not the result of the "mere" sensory observations of empiricism. In fact, although he begins "with the most familiar truths, with facts of hourly experience," which, when inductively built up, will result in "positions the most comprehensive and sublime," these comprehensive and sublime positions are not the end of the inquiry—it is *they* that "prepare the mind for the reception of specific knowledge" (1:449).

The passivity of "observations merely empirical" must be balanced, in scientific language recalling Wordsworth's, by the active engagement of the mind: "Method . . . must result from the due mean or balance between our passive impressions and the mind's own re-action on the same" (1:453). The basis of method, therefore, is the mind's search for relations among the passive impressions of the senses. "[W]here the habit of method is present and effective," Coleridge writes, "things the most remote and diverse in time, place, and outward circumstance, are brought into mental contiguity and succession" (1:455). Yet the science of method involves even more than this. Although "the RELATIONS of objects are prime *materials* of Method," the "indispensable condition of thinking methodically" is when mind enters in the "contemplation" of these relations (1:458), in the analysis of the relations of the relations.

Coleridge is insistent about the difference between method and classification. Method implies not just relation but progressive transition; it moves, as above, from "familiar truths" to "comprehensive and sublime positions" to "specific knowledge"—the collection of facts must follow, not precede, the idea that governs them. In the classificatory sciences, even in "the lowest attempt at a methodical arrangement . . . some *antecedent* must have been contributed by the mind itself; some purpose must be in view; or some question at least must have been proposed to nature, grounded, as all questions are, upon *some* idea of the answer . . ." (1:467). What Coleridge called "the common notion of Lord Bacon" in his 1819 philosophical lectures—"that you are to watch everything without having any reason for so doing"—is not method at all, for it lacks this mental antecedent.[28] Similarly, the alphabetical arrangements of dictionaries and encyclopedias cannot be called methodical, for they, too, present collections of facts without a governing idea and hence without progressive transition. They are "a mere dead arrangement" (1:457).

Preeminent for Coleridge among dead arrangements was the classificatory system of Linnaean botany. However important Linnaeus' system, it is merely a preliminary, a dictionary; unfortunately, many botanists, like Wordsworth's geologist, are content to classify and move on, ignoring the issue of relations and the even more important questions such relations raise:

> We acknowledge, we reverence the obligations of Botany to Linnaeus. . . . He invented a universal character for the language of Botany chargeable with no greater imperfections than are to be found in the alphabets of every particular language. As for the study of the ancients, so of the works of nature, an accidence and a dictionary are the first and indisputable requisites: and to the illustrious Swede, Botany is indebted for both. But neither was the central idea of vegetation itself, by the light of which we might have seen the collateral relations of the vegetable to the inorganic and to the animal world; nor the constitutive nature and inner necessity of sex itself, revealed to Linnaeus. Hence, as in all other cases where the master-light is missing, so in this: . . . what is BOTANY at the present hour? Little more than an enormous nomenclature; a huge catalogue, *bien arrangé*, yearly and monthly augmented, in various editions. . . . The terms system, method, science, are mere improprieties of courtesy, when applied to a mass enlarging by endless appositions, but without a nerve that oscillates, or a pulse that throbs, in sign of *growth* or inward sympathy. (1:466–69)

As part of the book of nature, the world of plants must be read, and Linnaeus' system is the dictionary and the grammar necessary for learning the language. But botany has not moved further, becoming, like a dictionary or an encyclopedia, a huge catalogue constantly added to and revised, but never with a sense of its relations to the rest of the organic and inorganic worlds. Significantly, Coleridge shifts metaphors to describe the organic wholeness that botany lacks; it is an inert mass, a dead arrangement without nerve or pulse, that does not grow as a living thing.

These complaints about classification and the emphasis on wholeness and relations and progressive transition do not occur in a vacuum, for the "Essay on Method" originally stood as the general introduction to, and organizing principle of, the *Encyclopaedia Metropolitana*, which Coleridge had been commissioned to edit. So annoyed was he by the changes made to his scheme by the publisher that he disassociated himself from the version of the "Essay on Method" that appeared in volume 1 of the *Metropolitana* in early 1818, and later that same year he incorporated much of the original version into the new edition of *The Friend*. As Richard Yeo has demonstrated, the *Metropolitana*, planned as a rival to the *Encyclopaedia Britannica*, deliberately eschewed the *Britannica*'s alphabetical arrangement in favor of a more methodical (in Coleridgean terms) system.[29] Although alphabetical arrangement had already become more popular than eighteenth-century "map of knowledge" approaches, Coleridge felt that the recent revolutions in knowledge—scientific and otherwise—demanded a new synthesis rather than what he saw as the *Britannica*'s abandonment of synthesis. The *Metropolitana* was therefore to emphasize the relations within the sciences and between the sciences and other branches of knowledge, the mental antecedent or idea on which each branch of knowledge was based, and the progressive nature of knowledge. While employing a traditional scheme, and even while using the language of Linnaean botany (class, order, genus, species) to organize his hierarchy of areas of study, Coleridge sought to make his classification a living rather than a dead arrangement.

In a footnote to his criticism of Linnaeus, Coleridge explains that "nature" can have two senses, the active/energetic (often referred to by Coleridge as *forma formans or natura naturans*) and the passive/material (*forma formata* or *natura naturata*), and that dead arrangement is the result of studying nature only in the latter sense (1:467n).[30] This material sense is "the sum total of all things, as far as they are objects of our senses, and consequently of possible experience—the aggregate of phaenomena, whether existing for our outward senses, or for our inner sense." By inner sense, Coleridge here means the imagination in its passive, involuntary operation as memory (see 1:177n), so, like Wordsworth, he considers the study of "things as they are" to be passive and inadequate. But Coleridge argues that this passive imagination is "more properly entitled Phaenomenology": when the mind is active in the contemplation of nature, it does more than exhibit things as they appear to exist to the senses and passions of the observer—it supplies "the inward principle of whatever is requisite for the reality of a thing" (1:467n). Yet, as is typical in Coleridge's polar thought, the study of nature in the active, energetic sense (what he calls the science of Dynamics) becomes complete only when reconciled with its "anion," this Phaenomenology. It is the reconciliation of Dynamics and Phaenomenology that for Coleridge constitutes Natural Philosophy.

Since method is based on the search for relations, the balance between passive sense impressions and the mind's re-action upon them, there correspond to the two senses of nature two kinds of relations: Law (or Idea) and Theory.[31] Of Law, Coleridge writes: "[I]n whatever science the relation of parts to each other and to the whole is predetermined by a truth originating in the *mind*, and not abstracted or generalized from observation of the parts, there we affirm the presence of a law, if we are speaking of the physical sciences, . . . or the presence of fundamental *ideas*, if our discourse be upon those sciences, the truths of which, as absolute truths, not merely have an independent *origin* in the mind, but continue to exist in and for the mind alone" (1:459). Law is generated not from phenomena but from the mind. This is in keeping with one of Coleridge's most basic criticisms of "the common notion" of Baconian induction, that the investigation of nature, the collection of facts, cannot be carried on methodically without "the prior assumption of some efficient law . . ." (1:467n). Phenomena of nature can confirm the validity of these mental truths, but they are not the foundation of them: the "remarkable fact" is that "the material world is found to obey the same laws as had been deduced independently from the reason" (1:462).

Theory, on the other hand, is constructed less from the reason and more from the material world. It is concerned with "the existing forms and qualities of objects, discovered by observation or experiment," and it attempts to bring unity to these forms and qualities, to "suggest the arrangement of many under one point of view" (1:463). The method built on Theory can thus only approximate the method built from Law (1:465). In addition to the danger that "arrangement" may merely facilitate, as in Linnaean botany, communication and recollection of facts rather than the understanding and control over nature that can lead to the discovery of Laws (1:463), Theory is necessarily limited and imperfect because it begins in generalization from particular facts, which, originating in observation and experiment, are themselves limited and imperfect (1:477). Indeed, according to Coleridge, the theory generalized from facts must itself be based on a theory: "what shall determine the mind to abstract or generalize one common point rather than another? and within what limits, from what number of individual objects, shall the generalization be made?" (1:476). Since sense impressions, unlike Laws, are in constant flux (1:462), many Theories are easily overthrown by the discovery of a single new fact (1:477). And Hypotheses are even more tenuous: a hypothesis is valid only when it is a "picture-language of an *idea* which is contained in it more or less clearly. . . . In all other instances, it is itself a real or supposed phaenomenon, and therefore a part of the problem which it is to solve. It may be among the foundation-stones of the edifice, but can never be the *ground*" (1:477–78). Theories and hypotheses are the "dress" of laws, and more often than not they serve only to "disguise and disfigure" them (1:478).

Examples may make these distinctions clearer. Like botany, zoology lacked a "master-light"; it was "weighed down and crushed . . . by the inordinate number and manifoldness of facts and phaenomena apparently separate, without evincing the least promise of systematizing itself by any inward combination, any vital interdependence of its parts" (1:473). This lack of system and "vital interdependence" (again note the metaphor of the living body)—in short, the lack of method—has led to premature assumptions divorced not only from Law and Idea but also from the evidence of phenomena. Coleridge hopes that the "objective truth" of a theory will be demonstrated by an "induction of facts . . . in nature," but such an induction of facts would still be secondary to "laws of organic nature," for "with the knowledge of LAW alone dwell Power and Prophecy, decisive Experiment, and lastly, a scientific method, . . . dissipating with its earliest rays the gnomes of hypothesis and the mists of theory . . ." (1:473). Stopping at facts, at *natura naturata* and the hypotheses and theories built upon them, leads us to worship our own abstractions.

In *The Friend*, Coleridge illustrates his contention that a perfect theory is impossible by taking up terrestrial magnetism. He quotes a writer in the Royal Institution's *Journal of Science and the Arts* to the effect that extensive global magnetic observations must be collected, collated, and "'brought into one focus'" before "'anything like a sound and stable Theory'" of terrestrial magnetism can be formulated (1:477n). This is "the common notion of Lord Bacon," and Coleridge undermines it. The writer, he says, is correct—a "sound and stable Theory" indeed cannot be formulated without the collection, collation, and focusing of material facts. But not only does this beg the question of how the facts are to be brought into one focus without first proposing the theory that the focusing is supposed to lead to, it leaves out "the necessity of a mental Initiative" (1:466), the connection to Law.

Similarly, though the phenomena of electrical attraction and repulsion were known for centuries, the study of electricity has advanced only with the development of the idea of polarity, which is what gives electrical science a method (1:478). Unfortunately, this idea has been clothed in many different theories of electricity, all "supported by insecure hypotheses" of fluids and chemical compounds and elementary matter, and all tending to obscure the truth. Yet the "theories and fictions of the electricians" nonetheless "contained an *idea*, . . . which has necessarily led to METHOD; implicit indeed, . . . but which requires little more than the dismission of the imagery to become constitutive," while the work of magnetists, despite a similar knowledge of basic facts, has "led to no idea, to no law, and consequently to no method" (1:479). Their hypotheses—that the earth itself is a giant magnet, or that a magnet is concealed within the earth—are devoid of an idea; they are "but repetitions of the same fact . . . , the *reiteration* of the problem, not its solution" (1:481).

For Coleridge, theory can sometimes lead to law, but the fact that theory originates in material objects and not in the mind makes the movement to law extraordinarily difficult. There is, however, yet another type of relation, aside from law and theory, that facilitates this movement, and it is the method of the fine arts (1:464). In a paragraph most interesting for its shifts in meaning, Coleridge originally assigns the method of the fine arts to theory, for the fine arts also "attempt to arrange the forms and qualities of objects determined from observation and experiment under one point of view." But there is too much "mental Initiative" in the fine arts to assign them exclusively to theory, so Coleridge amends his classification: the method of the fine arts also links theory and law. Yet when he considers all "that truly merits the name of Poetry in its most comprehensive sense," Coleridge finds that "there is a necessary predominance of the Ideas (i.e., of that which originates in the artist himself), and a comparative indifference of the materials." Poetry, as Wordsworth claimed, takes the objects of the material world and considers them not simply as they are, but as they appear to the poet's mind. For Coleridge, when "that which originates in the artist" predominates, the method of poetry—and in its "most comprehensive sense" poetry also includes science when Ideas are predominant—stands between the relations of law and theory, though closer to law. And in fact, Coleridge opens his discussion of method not with science but with literature: using Shakespeare as his example, he demonstrates "that union and interpenetration of the universal and the particular, which must pervade all works of decided genius and true science" (1:457), an echo of his claim in the philosophical lectures that "we must assume . . . as a ground of all reasoning a perpetual tendency at once to individualize and yet to universalize, or to keep *a balance*."[32] When he turns to the discussion of the sciences that I have been charting, he tells the reader he has already illustrated the truth of the "superior dignity" of Law in his remarks on Shakespeare (1:465–66).[33]

The method of the fine arts mediating between Law and Theory parallels the Imagination's mediating function between the Reason and the Understanding.[34] Like the *Biographia Literaria* and the *Statesman's Manual*, *The Friend* devotes considerable attention to the distinction between the Reason and the Understanding, "the which remaining obscure, all else will be as no system" (1:153). In this exposition, specifically designed to be "more popular" than his other attempts, Coleridge calls the Reason "the organ of the super-sensuous" (1:156), "the power by which we become possessed of principles, . . . and of Ideas, . . . as the ideas of a point, a line, a circle, in Mathematics; and of Justice, Holiness, Free-Will, &c. in Morals" (1:177n). The Reason bears "the same relation to spiritual objects . . . as the eye bears to material and contingent phaenomena" (1:155–56). Material and contingent phenomena—the impressions produced by the outward senses and the involuntary memory of past sensations according to the laws of associative psychology—are referred to by Coleridge as Sense and "comprise whatever is passive in our being" (1:177n). These impressions and sensations are the province of the Understanding, "the faculty by which we generalize and arrange the phaenomena of perception" (1:156). The Understanding, in other words, acts upon the evidence provided by the Sense, combining them into rules which the Reason in turn subordinates into "ABSOLUTE PRINCIPLES or necessary LAWS" (1:157). Demonstrating that the Laws of the Reason account for the facts of Experience is what, according to Coleridge, constitutes science in its widest sense as knowledge (1:157–58).

Arthur Lovejoy, in his classic discussion of Romanticism and Kantian transcendentalism, denies that the secondary position of the Understanding implied an anti-intellectual contempt for science: "In some of the uses made of [Kant's theory of knowledge], especially when it became popularized and was given application in various fields by men of letters and other non-philosophers, it sometimes did make for . . . a distrust of scientific methods as such. But though it became the fashion to speak condescendingly, and even contemptuously of the unfortunate *Verstand*, which admittedly was the 'organ' of scientific inquiry, this was not usually meant to imply that the Understanding is not all very well in its place."[35] As a philosopher as well as a man of letters, Coleridge, while subordinating the Understanding to the Reason, certainly did not distrust scientific methods as a whole, nor did he speak condescendingly or contemptuously of the Understanding. Thus, although his discussion of method is confined to its employment by the Understanding, Coleridge is always concerned with the relationship between the Reason and the Understanding, with trying to show why some methods (like botany's) cannot lead to the Laws, or confirm the Ideas, of Reason, while others (like electricity's) can.

The closeness of this relationship is evident in the fact that the Reason, when used in its more popular sense as the reasoning faculty of the mind rather than its specifically Kantian sense as the organ of the supersensuous, is the true scientific faculty, for it approaches the data of the Sense and the Notions of the Understanding *through* the Laws of the Reason. The example Coleridge uses to explicate his contention is both surprising and illuminating. He compares the "self-evident truth" that the radii of a circle are all equal, from which it is possible to deduce the properties of a lever considered as the spoke of a circle, to the instance in which a man, collecting and weighing the various applicable facts, determines "whether it would be better to plant a particular spot of ground with larch, or with Scotch fir, or with oak in preference to either" (1:158–59). "[A]ll men," says Coleridge, "have agreed to call the results of the first class the truths of science, such as not only are true, but which it is impossible to conceive otherwise: while the results of the second class are called *facts*, or things of *experience*. . . . Now when the mind is employed, as in the first case mentioned, I call it *Reasoning*, or the use of the pure Reason, but, in the second case, the *Understanding*. . . ." Like the similar examples of De Quincey and Hunt, Coleridge's example includes a deliberately homely image of a man deciding what kind of tree to plant. But unlike De Quincey and Hunt, Coleridge uses this as an example not of science but of "matter of fact"—when not linked to the Laws of the Reason, the Understanding merely assesses the various facts of experience and chooses the most likely explanation or course of action. This is naive Baconianism, reminiscent of Macaulay's example of the peasant who determines inductively that today's upset stomach is the result of yesterday's kidney pie. The "truths of science," on the other hand, are associated by Coleridge with the operation of the Reason. What is interesting, though, is that Coleridge uses as his example of science not merely the common one of geometry, but one from mechanics: the properties of a lever. The self-evident truths of geometry are indeed for Coleridge (as for virtually everyone else at this time) truths of which it is impossible even to conceive the negative, truths of the pure Reason. The principles of the lever, however, were presumably not originally deduced from the definition of the circle; although developed from facts of experience by the Understanding, the "essential properties" of the lever are reached by the Reason operating as "the scientific Faculty" on both the Notions of the Understanding and the Laws of geometry.

Coleridge's rather different conception of "matter of fact" is also evident in the *Biographia*, where unlike De Quincey and Hunt and Wordsworth he turns "matter of fact" against poetry—Wordsworth's poetry, no less. One of the weaknesses of Wordsworth's poetry, says Coleridge, apologizing for his use of a "new coined word," is an occasional "matter-of-factness" (2:126).[36] This matter-of-factness takes two forms: "a laborious minuteness and fidelity in the representation of objects, and their positions, as they appeared to the poet himself" (2:126) and "an apparent minute adherence to *matter-of-fact* in character and incidents; a *biographical* attention to probability, and an *anxiety* of explanation or retrospect" (2:129). In both cases, Coleridge complains about detail for detail's sake, but in the former he again contends for the insufficiency even of detail that involves the poet's mind. Poetry is more than the description of things as they appear to the poet.

Coleridge's criticism of Wordsworth also takes up the issue of knowledge and pleasure. According to Coleridge, Wordsworth, in his matter of factness, departs from the immediate aim of poetry, namely, the production of pleasure (2:130). Implicitly rejecting Wordsworth's replacement of the poetry/prose distinction with the more "philosophical" opposition of poetry and "matter of fact, or science," Coleridge argues, as he does in the *Biographia*, that poetry and prose (which includes science) can be distinguished *only* by their "immediate purpose"—poetry's is pleasure, prose's "the communication of truths" (2:12). Wordsworth's matter of factness in *The Excursion* is often noble, for it aids in the communication of truths, but in doing so it makes truth and not pleasure the immediate end of the poem. And where Wordsworth privileged pleasure as he yoked it to knowledge, Coleridge makes the link less absolute but more equal. Knowledge and pleasure are not necessarily connected for Coleridge, but they may be, and often are, in the best poetry and the best prose. The communication of truths—and we can see that Coleridge does not have in mind the sterile didacticism of a De Quinceyan literature of knowledge—may be attended by pleasure "of the highest and most permanent kind"; the communication of pleasure "ought" to have "truth, either moral or intellectual," as its "ultimate end" (2:12). Characteristically, where many other Romantics, including Wordsworth at times, tended to maintain and even widen distinctions, Coleridge seeks synthesis and reconciliation.

COLERIDGE AND BACON

Nowhere is the reconciliation of polar "opposites" more evident in Coleridge's thought than in the reconciliation of the Baconian and Platonic methods, a reconciliation long in Coleridge's mind, vital to his system of knowledge, and central to the "Essay on Method." Coleridge venerated the *Novum Organum*, once declaring that he believed "in my depth of being" that it was one of the three great works (along with Spinoza's *Ethics* and Kant's *Critique of the Pure Reason*) since the introduction of Christianity.[37] As early as 1803, sketching the plan of his own "Organum vere Organum," Coleridge said that his examination of the "Verulamian Logic" would include a "comparison of it with the Logic of Plato (in which I attempt to make it probable, that tho' considered by Bacon himself as the antithesis & Antidote of Plato, it is bona fide the same . . . )."[38] In a notebook entry on Plato, Coleridge expresses the wish that "I had the republication of the Novum Organum with a series of extracts from Plato in the Notes—I could pledge myself for the proof that whatever is true in the Baconian Logicè . . . is precisely Platonic."[39] His wish was nearly granted, for his friend Basil Montagu initially engaged Coleridge to translate and comment on the *Novum Organum* for Montagu's edition of Bacon's *Works*.[40]

The first reference to the reconciliation of Bacon and Plato in *The Friend* occurs in the important footnote about the active and passive meanings of "nature" and the definition of Idea. Having just asserted that "the very impulse to universalize any phaenomenon involves the prior assumption of some efficient law" which is "incapable of being abstracted or generalized from any number of phaenomena, because it is itself presupposed in each and all as their common ground and condition," Coleridge contends that "[s]uch is the doctrine of the Novum Organum of Lord Bacon, agreeing . . . in all essential points with the true doctrine of Plato, the apparent differences being for the greater part occasioned by the Grecian sage having applied his principles chiefly to the investigation of the mind, . . . and the English philosopher to the development of nature" (1:467n). The last part of the first section of the "Essay on Method" is devoted to a fuller demonstration of this essential agreement, of the "coinciding precepts of the Athenian Verulam and the British Plato" (1:488). If we separate "the *grounds* and essential *principles* of their philosophic systems" from specific points and practical applications, we find "that they are radically one and the same system: in that, namely, which is of universal and imperishable worth!—the science of Method, and the grounds and conditions of the science of Method" (1:487).

Reconciling two thinkers thought to epitomize "idealism" and "empiricism" was of course in keeping with Coleridge's polar philosophy, but to do so he had to show that *both* Plato and Bacon had been misinterpreted. Coleridge blames "the spirit that was at work in the latter half of the last century" for wrongly accusing the writings of Plato of "estranging the mind from sober experience and substantial *matter-of-fact*, and of debauching it by fictions and generalities" (1:482). Yet this critical late-eighteenth-century view of Plato was itself based in large measure on Bacon's own critical remarks, which Coleridge accounts for by saying that Bacon evidently had not read Plato but only his commentators (1:483, 487). In fact, says Coleridge, Plato's method is "inductive throughout": in his attacks on "the assumptions, abstractions, generalities, and verbal legerdemain of the sophists" he argues "not only *from*, but *in* and *by*, inductions of facts!" (1:482). Yet in making Plato an inductive thinker, Coleridge is careful to align him with a Baconianism much different from that created by the spirit of the second half of the eighteenth century.

Like the commentators who would follow him, Coleridge denied that the merits of Bacon's method lay in its specific recommendations:

> Nor let it be forgotten that the sunny side of Lord Bacon's character is to be found neither in his inductions, nor in the application of his own method to particular phaenomena, or particular classes of physical facts. . . . Nor is it to be found in his recommendation (which is wholly Independent of his inestimable principles of scientific method) of tabular collections of particulars. Let any unprejudiced naturalist turn to Lord Bacon's questions and proposals for the investigation of single problems . . . and put it to his conscience, whether any desirable end could be hoped from such a process; or inquire of his own experience, or historical recollections whether any important discovery was ever made in that way. For though Bacon never so far deviates from his own principles, as not to admonish the reader that the particulars are thus to be collected, only that by careful selection they may be concentrated into universals; yet so immense is their number, and so various and almost endless the relations in which each is to be separately considered, that the life of an ante-diluvian patriarch would be expended, and his strength and spirits have been wasted, in merely polling the votes, and long before he could commence the process of simplification, or have arrived in sight of the law which was to reward the toils of the over-tasked PSYCHE. (1:483–85)

Those features of Bacon's system usually taken as central to Baconian induction are for Coleridge not method at all. The collection and tabulation of facts is not only impractical but, lacking the active participation of the mind, doomed to fruitlessness. As he asserted in the published "Treatise on Method" in the *Metropolitana*, "[t]hose who talk superficially about Bacon's Philosophy" are in fact "nineteen-twentieths of those who talk about it at all."[41] The "true Baconian philosophy," on the other hand, consists in "a profound meditation on those laws which the pure reason in man reveals to him, with the confident anticipation and faith that to this will be found to correspond certain laws in nature."[42] Baconian science, properly understood and guided by the relational thinking that constitutes method, begins not in the senses but in the reason. This may sound more Coleridgean than Baconian, but Coleridge closes *The Friend*'s theoretical discussion of method by employing "our renowned countryman's own principles of Method" both to summarize the crucial components of the science of method and to "recapitulate the substance" of the broader transcendental doctrines "asserted and vindicated in the preceding pages" (1:488).

Unlike his followers, Bacon understood both the inadequacy of the senses and the necessary presupposition of Ideas in scientific inquiry: "Lord Bacon, equally with ourselves, demands what we have ventured to call the intellectual and mental initiative, as the motive and guide of every philosophical experiment; some well-grounded purpose, some distinct impression of the probable results, some self-consistent anticipation. . . . With him, therefore, as with us, an idea is an experiment proposed, an experiment is an idea realized" (1:489). The "organs of sense" can "apprehend" the "phaenomena evoked by the experiment," but the Reason, as the source of Ideas, is both what shapes the experiment and interprets its results (1:489–90). Coleridge associates the "pure and impersonal reason" with Bacon's "*lumen siccum*," the dry light of the intellect "freed from all the various *idols* enumerated by our great legislator of science . . . ; that is, freed from the limits, the passions, the prejudices, the peculiar habits of the human understanding, natural or acquired; but above all, pure from the arrogance, which leads man to take the forms and mechanism of his own mere reflective faculty, as the measure of nature and of Deity." The Baconian idols are for Coleridge creatures of the Understanding; in urging the elimination of the idols in scientific inquiry Bacon was urging reliance on the Ideas and Laws of the Reason rather than the Theories and Hypotheses chained exclusively to the Understanding and the Sense. Bacon teaches that the Understanding, because it is polluted by the idols, leads us away from the study of Ideas and Laws to its own subjective and idiosyncratic products: it "reflects the objects subjectively, that is, substitutes for the inherent laws and properties of the objects the relations which the objects bear to its own particular constitution" (1:492). In other words, if Theory (the relations of objects as seen through the

Understanding) is to lead to Law (the inherent relations of objects), the Understanding "requires the same corrections as the appearances transmitted by the outward senses."

This effort to correct the Sense and the Understanding, to eliminate idols in the search for Laws and Ideas of the Reason, is "the great object both of Plato's and of Lord Bacon's labours" (1:490). "[I]t will not surprise us," writes Coleridge,

> that Plato so often calls ideas LIVING LAWS, in which the mind has its whole true being and permanence; or that Bacon, vice versa, names the laws of nature, *ideas*; and represents what we have . . . called *facts* of *science* and central phaenomena, as signatures, impressions, and symbols of ideas. . . . [T]he discipline, by which the human mind is purified from its idols, and raised to the contemplation of Ideas, and thence to the secure and ever-progressive . . . investigation of truth and reality by scientific method, comprehends what [Plato] so highly extols under the title of Dialectic. According to Lord Bacon, as describing the same truth seen from the opposite point, and applied to natural philosophy, an idea would be defined as—An intuition or discovery of ideas of the divine mind, in the same way that they disclose themselves in things by their own signature, and this (as is proper to the dry light's Intellection) is not in sense perception. (1:492–93)

The Platonic Ideas are not airy abstractions but grounded in the objective Laws of the Reason, the Understanding freed from its idols. Baconian facts, similarly, are not mere objects of the senses but symbols of Ideas. Philosophy is "necessarily bipolar"; the only difference between Plato and Bacon is that Plato studies truth at the idea pole (the "science of intellect") and Bacon at the material pole (the "science of nature").[43]

Coleridge performs a similar reconciliation of Bacon and Kant, who has "completed and systematized what Lord Bacon had boldly designed and loosely sketched out in the miscellany of Aphorisms, his Novum Organum."[44] In his manuscript *Logic*, criticizing those writers who have "copied from copyists and repeated for the hundredth time the repeaters of Lord Bacon's assertions" (120), Coleridge both rehabilitates and reconstructs Bacon. Unlike his followers, Bacon understood the insignificance of common logic and the inadequacy of the senses (135–37). While the repeaters of Bacon's assertions babble about his specific methodological recommendations, Coleridge claims that "among all the works of modern philosophy, Bacon's *Novum Organum* has the immortal merit of having first fully and worthily set forth" the importance of the fact that "rules . . . should be *prevented* from being applied as principles to the determination of objective truth in real science . . . " (191, my emphasis). As in *The Friend*, Coleridge's interest is not in the rules but in the spirit of Bacon's philosophy, for it is the spirit of the *Novum Organum* that he can show to be in accord with Kantian logic. Logic is the science of the Understanding, but it must not remain severed from the Reason; in reconciling the immortal Verulam with the sage of Konigsberg, Coleridge hopes to bring the two poles together and reveal them as one.

In the *Biographia*, Coleridge recalls a time when he was not optimistic at all about the development and reconciliation of philosophical systems. His study of Locke, Berkeley, Leibniz, and Hartley for a time resigned him to the conclusion that "the sole practicable employment for the human mind was to observe, to collect, to classify" (1:141). In his despair, as De Quincey would do later, he contrasted Newton unfavorably to Milton and to Shakespeare, proclaiming that Newton's optical theories were the product of a passive and materialist mind.[45] But by *The Friend*, such contrasts are gone. Coleridge's list of "first-rate philosophers," those in whom "the imagination or preconstructive power have taken a scientific or philosophic direction," includes Kepler, Boyle, and Newton as well as Plato (1:416). As much "divinity of the intellect" was found in Shakespeare, Milton, and Bacon as in Plato, Demosthenes, and Homer (1:392). Yet one of the great philosophical weaknesses of his own day, according to Coleridge, was that Bacon and Newton were revered, Plato and Kant criticized or ignored, for the wrong reasons. "Every Platonist must revere Lord Bacon," Coleridge once wrote in his notebooks, "as the great Restorer of the genuine Platonic logic—viz.—Progress by Induction. The modern chemists talk of Bacon, but with the exception of Humphrey [*sic*] Davy, I know of none who have not as grossly offended against his Laws of Investigation in the one extreme, as the Schoolmen did in the other. The Schoolmen wished to find all things *in* their brains; the Priestley's to find everything *without* brains."[46]

Thinking dichotomously leads to the extremes represented towards the "Laws of Investigation" by Scholastic philosophy on the one hand, modern pseudo-Baconians on the other. Coleridge, of course, is interested not in the extremes but in their reconciliation, in the bridge built between subject and object, observer and observed, *natura naturans* and *natura naturata*, Reason and Sense (through the Understanding), Law and Facts (through Theory). Yet reconciliations cannot be made without making distinctions, and distinctions can often be taken, as they were by Hunt and De Quincey, for absolute differences. More directly and more openly than Wordsworth, Coleridge wanted to distinguish between those modern chemists who talk of Bacon and those rare exceptions, like Davy, who understand "the true Baconian philosophy." And so it is important for us to see why Wordsworth and Coleridge were both attracted and repelled by Davy's scientific methodology.

## DAVY AND THE SCIENTIFIC IMAGINATION

The career of Humphry Davy provides a useful counterpoint to those of Coleridge and Wordsworth. In his youth, Davy was torn between science and poetry as a vocation; his sudden appointment to the Royal Institution while still in his early twenties made the decision for him. As he became increasingly entrenched in the scientific establishment, eventually rising to the presidency of the Royal Society, Davy also became an embodiment of the methodological contradictions in the relationship of science and poetry. On the one hand, Davy was seen as an extraordinarily inventive and imaginative scientist; on the other, his invention of the safety lamp for miners was trumpeted as the epitome of the Baconian method. Despite his celebration of the virtues of Baconian induction in the traditional sense that Wordsworth and Coleridge loathed, he also significantly revised that method both in his writings about method and in his scientific practice. And although he tended over the years to privilege science over poetry, the very language he used for doing so often undermined his position.

A fragment of an early unfinished poem suggests the affinity Davy felt for both science and poetry, especially in the presence of nature:

> Thy awful height, Bolerium, is not loved
> By busy man; and no one wanders there
> Save he who follows Nature; he who seeks
> Amidst thy craigs and storm-beat rocks to find
> The marks of changes, teaching the great laws
> That raised the globe from chaos; or, he whose soul
> Is warm with fire poetic,—he who feels
> When Nature smiles in beauty, or sublime
> Rises in majesty.
>
> (1:38)

For the young Davy, nature is the inspiration and the haunt of both scientist and poet, of him who seeks the marks of the earth's changes and him whose soul burns with poetic fire. A similar connection appears in a poem about returning to a favorite childhood spot: "Here, through the trembling moonshine of the grove, / My earliest lays were wafted by the breeze,— / And here my kindling spirit learn'd to trace /

The mystic laws from whose high energy / The moving atoms, in eternal change, / Still rise to animation . . ." (1:59). Like the rugged mountain, this moonlit grove is a nurturer of scientific and poetic spirit, although this time there is a more specific, personal connection to Davy himself: the grove is the spot both where he composed "My earliest lays" and where "my kindling spirit learn'd to trace / The mystic laws" of moving atoms.

Even at this time, however, with a scientific career barely under way and with acquaintances among Bristol's literary community, Davy began to separate science and poetry, and move toward science. In a poem included by Southey in an edition of the *Annual Anthology*, Davy chronicles his movement from "the fairy fancy [who] ruled my mind" to "awful reason," who "claims my last, my loftiest song; / She leads a brighter maid along, / Divine Philosophy" (1:76–77). In a variation of a common Romantic motif, Davy celebrates in poetry his decision not to pursue poetry as a vocation, but just in case the point was lost on his mother, in the letter to her that accompanied a copy of this first published poem, Davy wrote: "do not suppose I am turned poet. Philosophy, chemistry, and medicine are my profession" (1:71).

Moving to London and his post at the Royal Institution, Davy quickly discovered that "fairy fancy" and "Divine Philosophy" could not be so easily sundered. The young man who had yoked the two with enthusiasm found, on choosing one as his vocation, that his position demanded, and his predilections accepted, some measure of accommodation between them. His public scientific lectures required a well-rehearsed dramatic flair that led him to defend his efforts "to excite feelings of interest" rather than "to give minute information" (2:310). For Davies Gilbert, Davy's successor as president of the Royal Society, "the poetic bent of Davy's mind" was the source of "the distinguishing feature in his character, and in his discoveries,—a vivid imagination sketching out new facts in regions unexplored."[47] But this "poetic bent" and "vivid imagination" often clashed with his tendency to claim, especially as he grew older, that the path he had chosen was in fact the more valuable, more useful, more enduring of the two. In his final work, the posthumously published philosophic dialogues *Consolations in Travel; or, The Last Days of a Philosopher,* Davy claimed that the proper form of scientific presentation is sparse and clear, almost unpoetic. The main speaker asserts that "[i]n detailing the results of his experiments, and in giving them to the world, the chemical philosopher should adopt the simplest style and manner; he will avoid all ornaments, as something injurious to his subject" (9:366). Yet, characteristically for Davy, even this strongly worded condemnation of "ornaments" is not consistent. Not only does it contain a bit of narrative the purpose of which is to illustrate the rules for scientific presentation (much as Davy's experiments illustrated the laws of natural science presented in his public lectures), but it also occurs within a series of dialogues imaginatively conceived and poetically presented, dialogues which often discuss experiments in what is very much not "the simplest style and manner."

This point is illustrated by studying two uses of one of Davy's favorite metaphors to illustrate the differences between science and poetry. In an early undated notebook entry in which he compares the "exertion of talent" required in works of science with that in works of the imagination, Davy insists that it is difficult to differentiate between them. But imagination is "*merely* the vivid but *vague* association of images with passion," while reason, the governing faculty of physical science, associates images "according to facts observed in nature" (1:148; my emphasis). Wordsworth and Coleridge, of course, would have agreed that the poetic imagination associates natural objects with the passions of the poet's mind, but not that this association is vague. Nor would they have agreed with the dichotomy implied by Davy's elaboration of this claim: "The power of the mind, in the fervour of poetical composition, flows like a mountain torrent—sparkling, foaming, beautiful and grand; but passing principally over rocks, and nourishing only the solitary tree, or the flowers of its mossy borders. The energy of the understanding employed upon the truths of nature has a calm and quiet progress; in its motion it is like the navigable river; it bears upon it ships; it waters a fertile country; and what it wants in beauty it possesses in benefit; what is deficient in rapidity is supplied in strength." The description of the poetic imagination is itself imaginative: the language of the simile is vivid and the diction "poetic." "Sparkling, foaming, beautiful and grand" onomatopoetically recreates, in its broken series of adjectives full of liquid consonants, the sound and movement of the mountain torrent, which in turn represents "the fervour of poetical composition." Lost in the fervor of poetical composition, however, is the connection between pleasure and knowledge—Coleridge's distinction between the immediate aims of poetry and science is here implicitly raised to a difference in ultimate aims. Yet Davy's description of the operation of the reason in scientific pursuits is also highly imaginative. Having begun the simile about the poetic faculty, Davy extends it to the reason. Scientific thought moves slowly and calmly like a wide river. The short phrases of the first part, which are connected with ors and buts to reflect the rapid, shifting movement of the stream, give way in the second part to the series of longer, parallel expressions ("it is like . . . , it bears . . . , it waters . . ."; "what it wants in beauty it possesses in benefit; what is deficient in rapidity is supplied in strength") which depict the "calm and quiet progress," the stately order, of both the river and the understanding. Characteristically, even as he sets up this science/poetry, reason/imagination dichotomy most forcefully, Davy also talks about science most poetically, blurring the distinction he is trying to enforce.

Near the end of an 1808 lecture, however, Davy employs this simile in a different way. He defends "experimental science" from the charge that its "truths" lack "practical application" by equating science with "poetry and the fine arts" and arguing that anyone who questions the utility of a poem would not be thought worthy of a response (*Remains* 58). This curious defense is quickly abandoned, as Davy goes on to claim that science is in fact superior to poetry because its discoveries *do* have practical applications. "Like a stream rising amongst mountains," says Davy, "[natural science] flows from the heavens to the earth, and though in its beginnings it is known only to a few, its benefits gradually spread and at length it adorns and fertilizes an extensive and populous district" (*Remains* 60). This time, science is navigable river *and* mountain torrent. In its utility it no longer lacks beauty (the river "adorns" as well as "fertilizes"), and its origins include the quasi-divine status accorded to muse-inspired poetry.

The movement here is from a very negative formulation of science's utility (we value poetry despite its lack of utility, so why not value science, which is no more "useless" than poetry), a formulation in which poetry is presented as the established pursuit against which science must justify its credentials, to a much more positive one. Since science does have practical applications, it is implicitly able to displace poetry from its position of authority; science can do all that poetry can do, and more. The strategy is similar in the *Consolations*, where the principal speaker likens the pleasure of pure "chemical philosophy" to the pleasure normally associated with the contemplation of nature in poetry, but then turns Wordsworth's notion of science's need for transfiguration on its head: "There is no absolute utility in poetry; but it gives pleasure and refines and exalts the mind. Philosophical pursuits have likewise a noble and independent use of this kind; and there is a double reason for pursuing them, for, whilst in their sublime speculations they reach to the heavens, in their applications they belong to the earth" (9:360). In Davy's scheme, it is poetry that, despite its divine exaltation of the mind, needs to be transfigured and brought to earth in human form. Pleasure without utility is inferior to pleasure and utility.

This reference to the "sublime speculations" of science is not simply the effusion of a dying man in a partially mystical work, but part of an effort to establish an analogy between the canons of scientific discovery and those of aesthetics. As early as 1807, in an article entitled "Parallels between Art and Science," Davy links the truths of the natural sciences and the fine arts, saying that Newton, Shakespeare, Michelangelo, and Handel share a similar genius, for the philosophical mind must employ imagination as well as reason. "Discrimination and delicacy of sensation, so important in physical research, are other words for taste; and the love of nature is the same passion, as the love of the significant, the sublime, and the beautiful" (8:307). Once again, science is compared to the standard which is poetry's aesthetics, with conceptions of taste, the sublime, and the beautiful. But, if the pleasure of science does not quite match poetic pleasure, and therefore makes science less popular than poetry, science enjoys the more decisive advantage of being more durable. Works of art decay, but those of science remain, able to speak their universal truths directly to those of other countries and other times: "Nature cannot decay; the language of her interpreters will be the same in all times. It will be an universal tongue, speaking to all countries and all ages . . ." (8:308). The connection of Newton and Shakespeare as imaginative and philosophic thinkers was, as we have seen, shared by Wordsworth and Coleridge, but the contrast with De Quincey's view of the language of the interpreters of nature could not be more striking. For De Quincey, the literature of knowledge, including Newton's *Principia*, was spoken with not a universal but an ephemeral tongue, forgotten when revised or refuted or superseded, while the literature of power, of Shakespeare and Milton, truly spoke "to all countries and all ages."

Yet Davy's sense of the permanence of scientific truths elsewhere leads him to elevate Newton and Bacon above Shakespeare. In the same notebook entry where he likens the poetic imagination to a mountain torrent and the scientific imagination to a navigable river, Davy makes clear which he favors: "At that time when Bacon created a new world of intellect, and Shakespeare a new world of imagination, it is not a question to me which has produced the greatest effect upon the progress of society—Shakespeare or Bacon, Milton or Newton. . . . [T]he influence of [Shakespeare's] wonderful works is limited by the pleasure that they give" (1:147). Wordsworth concurs in the preface to the *Lyrical Ballads* that the poet's "one restriction" is "the necessity of giving immediate pleasure" (1:139), but he argues that the poet and the man of science share this restriction—indeed, that without pleasure there is no knowledge. For Davy, on the contrary, Milton and Shakespeare are not truly international, are unable to truly transcend their culture, because theirs is not a universal language, as the language of science is: "In natural science there is one language universally intelligible,—the language of facts; . . . it is permanent as the objects of nature; it is the same in the city of Paris and of London. Whenever the name of Newton is pronounced, it is pronounced with reverence" (1:148).

On the surface, this claim seems vastly at odds with Wordsworth's notion that "the language of facts" is actually a very limited base on which poetic language is built. It is evidence of how scientists could invoke the "facts" of the Baconian/Newtonian tradition against the vague and emotional language of poetic pleasure, and it certainly helps to explain why the Romantics reacted so strenuously against the threat such invocations could pose to poetry even as they incorporated elements of Baconian/Newtonian facts into their poetics. Yet without diminishing Davy's telling and characteristic effort to privilege the universality of the discourse of natural science, it is also important to place it in its more limited context. As in Wordsworth's conception that poetry becomes poetry only when the "facts" generated by Observation and Description have been transformed by the other poetic powers, Davy acknowledges that there is more to science than this language of facts: the imagination, which is the source of "conjectural inferences," is "a noble instrument of discovery" (8:317). Uncontrolled, it can lead to "premature . . . theory" and "hasty generalizations," but if used properly, it can assist in the development of useful hypotheses and valid theories.

## DAVY AND THE BACONIAN HERITAGE: THE POETRY OF SCIENTIFIC INDUCTION

The friendship of Wordsworth and Coleridge with Davy, and the poets' enthusiasm for their friend's scientific investigations, suggest that they saw Davy as a sharer in their veneration for Bacon and Newton and their disdain for eighteenth-century Baconians and Newtonians. But Wordsworth's uneasiness over some of the remarks in Davy's "Introductory Discourse"—remarks no doubt reinforced by some of Davy's later pronouncements—and Coleridge's dismay over Davy's acceptance of atomism, suggest that Davy could at times himself come across as a reverer of "the common notion of Lord Bacon." Like Wordsworth, Davy was not systematic in developing his position on the methodological tradition of British science, but his views are perhaps even more shifting, often studded with qualifications and contradictions.

In part these tensions are the result of the circumstances of Davy's life. On the one hand, Davy was an English scientist attempting to establish himself and his areas of interest within the Baconian/ Newtonian tradition. With his success, he found himself among a social and professional elite: knighted, married to an aristocrat, and later the president of the Royal Society. On the other hand, especially early in his career, Davy was associated with lesser known, sometimes controversial establishments, first with Beddoes' Pneumatic Institute and then with the newly formed Royal Institution. An inheritor of Newton's conceptual genius but not his mathematical ability, Davy was not a Cambridge don or gentleman philosopher, but a self-made, full-time scientist whose income was generated in large part by his ability to promote his science and himself in public lectures. Thus there is constant evidence in his writings of the desire to locate himself both within and without the methodological heritage of British science.

In an 1808 lecture, he refers to "the legitimate practice" for carrying out science as "that sanctioned by the precepts of Bacon and the examples of Newton . . . to proceed from particular instances to general ones, and to found hypotheses on facts to be rejected or adopted according as they are contradictory or conformable to new discoveries" (8:276). This is the standard genealogy, slightly modified only in its inclusion of hypotheses; the modification is not unimportant in the face of "hypotheses non fingo," but the emphasis is still on the control and limitation of hypotheses, which must be grounded in facts and subject to verification. An even more traditional attitude is expressed in his textbook on chemistry, *The Elements of Chemical Philosophy* (1812), when Davy expands his presentation of Bacon:

> This illustrious man . . . was a still greater benefactor to the science, by his development of the general system for improving natural knowledge. Till his time there had been no distinct views concerning the art of experiment and observation. . . . He taught that Man was . . . capable of discovering truth in no other way but by observing and imitating her [Nature's] operations; that facts were to be collected and not speculations formed; and that the materials for the foundations of the true system of knowledge were to be discovered, not in metaphysical theories, not in the fancies of men, but in the visible and tangible external world. (4:15–16)

The collection of facts in the visible and tangible external world is preferred to the forming of speculations and metaphysical theories that exist only in men's fancies. The venues of these remarks—an introductory public lecture to a popular audience and a basic text in chemistry—are significant in that the comments both reflect and contribute to the "common view" of Bacon as the promulgator of "the true system of knowledge."

Yet when addressing his colleagues, Davy is more bold. In his 1820 presidential address to the Royal Society, that bastion of the Baconian/Newtonian tradition, Davy eulogizes "our great masters, Bacon and Newton, [and] that sober and cautious method of inductive reasoning," but he also goes on to urge his fellow scientists to utilize hypotheses carefully "as leading to the research after facts"—as part of the scaffolding of the building of science (7:14). The shift is subtle but important: hypotheses are not now *founded* on facts but *lead to* the research after (i.e., the research which seeks to establish) facts. Hypotheses do not simply emerge from disinterested observation and the collection of facts but must be formulated, at least provisionally, before we can set up experiments or carry out observations. Davy wanted the freedom to formulate and apply hypotheses playfully; hypotheses, he once wrote in his notebooks, "should be formed with rapidity, applied with ease, and eternally varied; they should be the instruments of thought" (1:152). Yet these "secret amusements of the mind," these "phantasmagoria representations of the intellect," should be hidden until tested and verified, and discarded if found to be erroneous or inadequate.

Davy's terminology is not Coleridge's—he uses "hypothesis" in a much looser sense—but it is easy to see what is either Coleridge's influence on Davy's thought or Davy's congeniality with Coleridge's thought (or, more likely, both). Davy stresses the need in scientific investigation for what Coleridge calls "mental initiative," for the active participation of the mind in gathering facts, formulating hypotheses, and setting up experiments. Without this mental initiative, there is mere Coleridgean hypothesizing or what Davy calls "hasty generalization." Yet this mental initiative must also have some basis in physical phenomena and be subject to experimental verification. When it is not, scientists cling overzealously to the creatures of their own brains, the Baconian idols. In the 1810 "Lecture on Electro-Chemical Science," Davy quotes approvingly Bacon's complaint that "Men . . . are continually carrying too far their own favorite theories," but he goes on to castigate Bacon for the same failure (8:258).

Crucial to the mental initiative of Davy's scientific methodology is the use of analogy. Although analogy is obviously not a concept that originates in, or is exclusively connected with, scientific thought, its association with inductive reasoning is a strong one during this period. In Johnson's *Dictionary* (1758) we find no scientific usage for analogy, but by the *Century Dictionary* (1889), one of the definitions of analogy is that of a form of reasoning involving inference and therefore similar to induction. Under the definition of analogy as "equivalency" or "likeness of relations," the *OED* offers examples from the sixteenth century, but most of its nineteenth-century examples come from scientists. As a method of logical reasoning, analogy appears at least as early as 1602, but it is the example from Mill's *Logic* that first connects it specifically to scientific induction. The only definition of analogy in a specifically scientific sense is related to natural history, in the way that botany, zoology, comparative anatomy, etc., compare the physical characteristics and/or functions of one organism with those of another. The first example cited here is Davy himself, from the 1814 *Agricultural Chemistry*. And Foucault explicitly associates analogical thinking in the classificatory sciences with the late-eighteenth-century shift from visible structure to the more abstract concept of function as the organizing principle of these sciences.[48]

For Davy, the ability to think analogically is arguably the key to scientific success; it permits the scientist to connect apparently disparate phenomena, often in apparently disparate areas of study. Analogy is thus easily associated with the more freethinking, speculative aspect of scientific method, but even in this role it is essential to the process which makes sense out of an inchoate mass of experimental observations. Davy can call speculation an untenable process of making analogies based on words rather than facts, while still claiming that it is (with observation and experiment) one of the three "foundations of chemical philosophy" (4:2). Indeed, in developing his analogy between natural science and the human body and soul, Davy argues that facts are the "body" of natural science but that analogy is its "governing spirit" (8:167). His own scientific work is full of examples of this ability to reason by analogy, particularly in his pioneering work in connecting electrical and chemical phenomena.

Such successfully demonstrated analogies reinforce the assumption that the world is, for all its flux, ordered and interconnected, an assumption underlying both Davy's efforts to cross scientific fields and Coleridge's desire to combine all branches of human knowledge into one coherent system. More importantly for Coleridge, since the science of method is the study of relations, and the relations of "things the most remote and diverse in time, place, and outward circumstance" (1:455), thinking analogically is the essence of method. The "first charm" of

Davy's chemistry is thus for Coleridge the "ability to link apparently unlike things" (1:471). But linking apparently unlike things requires the ability to see differences and make distinctions as well as to formulate relations, to individualize and to generalize. The Understanding, which functions through "individualization, in outlines and *differencings* by quantity, quality and relation" (1:515), combines the "multifarious impressions" of Sense into "individual notions," and then develops these notions into rules "according to the analogy of its former notices" (1:157). Its generalizing capacities are limited to generalizing from phenomena, a process which cannot lead to Theory and Law without some stronger link to the mind. Thus the mark of the "methodizing intellect," whether a Hamlet's or a Davy's, is the "disposition to generalize" (1:452), the ability to apply the mental initiative of the Reason to the differences and distinctions of the Sense and the Understanding, and find relations.

Analogy fails for Coleridge when it does not reconcile distinction and relation but concentrates on one or the other. The failure of natural theology (especially its Paleyan version), which of course depended heavily on the analogy between design in the human and natural worlds, lies in its inability to remember that analogy implies distinction as well as similarity. The "enlightened naturalist," without denying that the resolution of his investigation of efficient causes into final causes may be the ultimate aim of all philosophy, "resists the substitution of the latter for the former as premature, presumptuous, and preclusive of all science" (1:498). The "presumption of a something analogous to the causality of the human will" in nature is valid for distinguishing nature's agency from "a blind and lifeless mechanism," but not for "assigning to nature, as nature, a conscious purpose." In the "meditative observation of a DAVY," chemistry also attempts to avoid seeing nature as a blind and lifeless mechanism, but its burden is the need to reconcile rather than distinguish. Even as it isolates the elements into individual substances, chemistry must be guided by a mental initiative that shows "the unity of principle through all the diversity of forms," as diamond and charcoal, both pure carbon, "are convoked and fraternized by the theory of the chemist" (1:470–71). The Priestleys and the Daltons, obsessed with mechanistic fictions such as atoms and lacking a sense of the Law of Polarity, try to "find everything without brains."

Yet there are also tensions lurking beneath these broad similarities in the methods of Coleridge and Davy. Foremost among them is Davy's sense of hypothesizing as a guilt-ridden activity that should be carried out covertly. Although essential, it is something to be done in secret and brought to light only when proven not to be "phantasmagoric." Coleridge, of course, feels no such shame about his own mental initiatives, but he is also less sanguine than Davy about the prospects of a Law evolving from Theory and Hypothesis. While Davy shares Coleridge's interpretation of Bacon as a critic of reliance on sensory data, he does not share Coleridge's view that experiment and theory, as constructions of the Understanding, also require correction. For Davy, experiment is relatively straightforward: "active inquiry by the trial of results, seems to be that which is best fitted to our nature, and to the development of truth" (*Remains* 163). Baconian induction, modified to provide greater latitude in the use of the imagination to formulate hypotheses, is not only the best method, but the most natural method.

These tensions involved in recasting the methodological tradition of Bacon and Newton are epitomized in the work which Davy saw as his monument to the inductive method: the development of the safety lamp for miners. If his lectures at the Royal Institution made him a celebrity in London, it was the invention of the safety lamp that cemented Davy's position as a national hero. It was a marriage of pure and applied science, of knowledge and utility, of metropolitan learning and provincial industry—a marriage signaled by the title of Davy's account of the invention: "On the Safety Lamp for Preventing Explosions in Mines, Houses Lighted by Gas, Spirit Warehouses, or Magazines in Ships, Etc., With Some Researches on Flame." The investigation of an eminently practical problem not only has other utilitarian applications, but is also inseparable from theoretical "Researches on Flame." The mine owners who commissioned the invention told Davy that the discovery could, by itself, make his name immortal. Joseph Banks, then president of the Royal Society, wrote Davy that he had placed both science in general and the Royal Society in particular on better public footing by showing how the application of natural philosophy could eradicate a social evil. Davy himself saw his research as a model, both for the applicability of pure science and for the proper employment of the inductive method. He told the mine owners that "my success . . . must be attributed to my having followed the path of experiment and induction" (1:208), and he wrote in his account of the discovery that his research details "the gradual progress of inquiry in which every step was furnished by experiments or induction" (6:4). The invention of the safety lamp, in other words, is as much a monument to the powers of induction from a perspective of utility as Newton's theory of gravitation was from a perspective of pure knowledge.

Davy's inductive case study does not, however, recount all the steps leading to the invention. Davy acknowledges passing over erroneous hypotheses and unsuccessful or inconclusive experiments, saying that "[i]t will be unnecessary to dwell upon preliminary and unsuccessful attempts, and I shall proceed to describe the origin and progress of those investigations which led me to the discovery . . ." (6:10). What Davy suppresses is the un-Baconian element of his procedure, the somewhat unsystematic patchwork of guesses, speculations, postulated analogies—in short, much of the imaginative aspect of the process. While allowing his intellect free play, he keeps out of sight "the secret amusements of the mind"—unless they lead directly to a truth. Instead, at each successive step in the inquiry, a problem is formulated, the possible answers tested, and the best solution adopted. For example, once Davy determines that the lamp's flame should be housed in a cage of wire gauze, the question becomes what size the holes in the gauze should be. Davy then performs a series of experiments with different sizes of holes and different thicknesses of wire, selecting the one that performed the best in his tests. Each step is made certain and secure before proceeding to the next. The "preliminary and unsuccessful attempts" are a vital part of his actual method, but they undermine this Baconian process of gradual and successive induction. Only when they are deleted from the public account can induction be presented as leading directly and inevitably to Truth, and thus eventually to Utility.

This tendency to suppress or subordinate the poetic in favor of the scientific, to trumpet in public naive inductivism rather than his "vivid imagination," opened possibilities in the interpretation of Davy's method that often made Wordsworth and Coleridge uneasy. When Henry Brougham reviewed Davy's Bakerian Lectures on electrochemistry for the *Edinburgh Review,* it confirmed for Coleridge the worst possible reception of the work that he saw extending the law of polarity into chemistry. While Brougham had high praise for the significance of Davy's researches, he also consciously diminished any sense of Davy's genius and imagination, turning him instead into an industrious Baconian fact collector. Davy, in Brougham's eyes, was at the right place at the right time, enabled by the facilities and financial largesse of the Royal Institution to obtain the enormous galvanic battery on which he could follow through on the hints and implications in the work of other, more original, minds. Moreover, Brougham simply ignored the products of Davy's own original mind, though it is clear from his comments that, as in the later paper on the safety lamp, the structure of the lectures and of Davy's language permitted such emphasis on facts over theory:

> Satisfied that the experimental investigation is the most material part of the work, . . . and that the facts already within our reach are insufficient for the foundation of a general theory, we have deemed it proper to confine our attention almost exclusively to a history of the subject, . . . and, without entering into any discussion of the hypotheses struck out by Mr. Davy, or even of the inferences which he is entitled to draw, we have reserved for a more mature branch of the Inquiry, whatever we may have to deliver on these heads. In so doing, we have indeed only followed our author's own example; for nothing is more praiseworthy in his treatise, than the caution and modesty with which he ventures to suggest, rather than lay down, his theoretical opinions; and he uniformly keeps them in the background, applying himself almost exclusively to the multiplication of facts, and repeatedly admitting that the time for theorizing is not yet come. Even at present, however, . . . we may be permitted to express the delight which we have received from his strict and patient induction.[49]

Coleridge was furious: "Had the Author had the Truth before his Eyes," he wrote to Davy, "and purposely written in diametrical opposition, he could not have succeeded better."[50] A year later, Coleridge defended Davy in *The Friend*, but the decision to retain that defense in the 1818 version even after the cooling in their friendship suggests that Coleridge saw the passage as a vindication of his own sense of method against the more "common" Baconianism of men like Brougham. Coleridge turned what Brougham called Davy's "strict and patient induction" inside out, emphasizing instead the very hypotheses and inferences that Brougham had glossed over. Davy's discoveries were "preconcerted by meditation, and evolved out of his own intellect" (1:530). His experiments were conducted in what Coleridge considered the true Baconian fashion, as a means of testing the Ideas of the Reason, "for the purpose of ensuring the testimony of experience to his principles, and in order to bind down material nature under the inquisition of reason, and force from her . . . unequivocal answers to *prepared* and *preconceived* questions" (1:530–31). Yet the very different responses of Brougham and Coleridge to Davy's electrochemical researches highlight the fundamental tension in Davy's work: although he helped to redefine scientific methodology in such a way as to open a place for the imagination and to appropriate the poetic faculty for science, his public Baconianism and his desire to privilege science over poetry even as he incorporated poetry into his work further contributed to Romantic uneasiness with that methodological tradition. If Wordsworth was critiquing a conception of science and scientific method already being challenged from within the scientific community itself, Davy could often seem blind to the degree to which Romantic poetics had absorbed the language and methodology of science.

The cases of Wordsworth, Coleridge, and Davy reveal a spectrum of positions on the relationship of science and poetry, but within that spectrum is a consensus on important methodological points. This has made it possible both for the Victorians and for us to construct very different readings of the Romantic attitude toward science, depending on whether the consensus or the spectrum is emphasized. Yet it is important to see that the methodologies of both science and poetry were being reshaped during this period, and that both scientists and poets were actively involved in this process. Romantic critiques of science played a role in the reinterpretation of scientific methodology, while scientific critiques of poetry contributed to and influenced Romantic poetics. Poets and scientists seem generally to have been aware of this mutual interpenetration, but each group was, especially in its rhetorical flourishes, capable of caricaturing the other, usually for the purpose of creating that cultural schizophrenia that could be so conveniently exploited.

Yet, to come back to George Dodd's comment in *Nature,* it seems that today there is a much more monolithic conception of the poet than of the scientist. Romanticism has left us with a powerful vision of the artist as solitary individual, sitting alone in a garret and singing a passionate and deeply individual song. Our image of the scientist, however, is more contradictory. On the one hand he is an Einstein, an inspired genius. On the other hand he is a faceless, emotionless automaton, a plodding collector of facts. The latter image testifies both to the lingering power of the rhetoric that helped to establish the authority of scientific discourse and to the continued Romantic hostility toward that seemingly "inhuman" aspect of science. If an Einstein undercuts that image, it is the exception that proves the rule: we call him poetic, and we assert that even if science and poetry need not always be sundered, they usually are. In the same fashion, the scientist hears that Wordsworth took an interest in science and is pleased to know that such rapprochement can occur, but his surprise at this violation of the normal order of things subtly reconfirms that poets and scientists have little to say to each other. As we deny the existence, or at least the inevitability, of two cultures, we too often do so in language that assumes the dichotomies we are trying to subvert.

In part this is because the dichotomies are our creation. As I have already hinted, our cultural impressions of scientist and poet, however deeply steeped in Romantic attitudes, also misread those attitudes: for Wordsworth, the poet is a communal figure, the scientist—even the good scientist, who remains connected to mind, to living nature, to the rest of humanity—is a lonely, isolated figure. The famous lines in *The Prelude* about the statue of Newton at Trinity ("Newton with his prism and silent face, / The marble index of a mind for ever / Voyaging through strange seas of Thought, alone"), in confirming this, also suggest the need both to recover these complicated Romantic attitudes and to uncover our own readings and misreadings of them. I have tried in this chapter to restore some of the complexities and tensions of the Romantics' response to science and its methodology, for those complexities and tensions are evident in my readings of Victorian texts that appropriate scientific language, theory, and methodology. But this restoration is also necessary if we are to expose the assumptions about the disjunction of science and literature that remain inherent in current critical discussions purporting to explode that disjunction—and which, to be fair, are difficult to escape from, even in my own discussions. We should not be surprised to find the poetic and the scientific intertwined, in other words, because poetry and science were fundamentally intertwined in the poetry and science of the Romantic era, an era whose views on these two ways of expressing facts continue to shape, however selectively and however unconsciously, our own.

# 3

# Seeing through Lyell's Eyes

## The Uniformitarian Imagination and *The Voyage of the Beagle*

In his *Preliminary Discourse on the Study of Natural Philosophy* (1830), John Herschel writes of the influence the study of the physical sciences has had on the study of the human sciences:

> The successful results of our experiments and reasonings in natural philosophy, and the incalculable advantages which experience, systematically consulted and dispassionately reasoned on, has conferred in matters purely physical, tend of necessity to impress something of the well weighed and progressive character of science on the more complicated conduct of our social and moral relations. It is thus that legislation and politics become gradually regarded as experimental sciences; and history, not, as formerly, the mere record of tyrannies and slaughters, . . . but as the archive of experiments, successful and unsuccessful, gradually accumulating towards the solution of the grand problem—how the advantages of government are to be secured with the least possible inconvenience to the governed.[1]

Herschel's view that the method of the physical sciences can be applied to the study of the past, that history is not just a sequence of events but an "archive of experiments," was not new. What is new, however, is Herschel's claim that this development of scientific history has changed the historian's sense of the content and the pace of history. No longer is history seen as a succession of violent political events, a "mere record" of "tyrannies and slaughters" packed closely together, but as gradually developing and accumulating experiments in government.

Earlier in that same year, the first volume of Charles Lyell's *Principles of Geology* also appeared. Lyell believed that the time had come for the study of the earth's history to abandon its reliance on hypothetical forces of tremendous violence and scope operating over short periods of time in favor of the gradual accumulation of less powerful, local forces (like erosion) operating over long periods. To illustrate the difference between his own approach and that of his opponents, Lyell drew an analogy with the study of history:

> How fatal every errour as to the quantity of time must prove to the introduction of rational views concerning the state of things in former ages, may be conceived by supposing that the annals of the civil and military transactions of a great nation were perused under the impression that they occurred in a period of one hundred instead of two thousand years. Such a portion of history would immediately assume the air of a romance; the events would seem devoid of credibility, and inconsistent with the present course of human affairs. A crowd of incidents would follow each other in thick succession. Armies and fleets would appear to be assembled only to be destroyed, and cities built merely to fall in ruins.[2]

Lyell's view of history merges with Herschel's: when history focuses on tyrannies and slaughters, the destruction of armies and the fall of cities, it ceases to be dispassionate and scientific and turns instead into a tale of romance. When geology relies on forces similarly sudden and violent, it produces a narrative of the earth's history that is similarly fanciful and romantic. Lyell's approach transforms catastrophic "romance" into a narrative of realism, but this realism is itself obtained through an imaginative process that accounts for even the most violent changes as effects of gradually operating forces.

Lyell's book, with its deliberate challenge of widely held views, threw the geological community, and the cultural community at large, into a methodological debate about the techniques appropriate for reconstructing the history of the earth. Between 1830 and 1872, Lyell's *Principles*—the geological bible for those who accepted what soon came to be christened "uniformitarianism"—went through eleven editions, its content revised by Lyell to keep up with this debate and with the changes in his own thought. But this controversy in geological methodology was of course also connected to the more general interest in scientific method as a whole. In this chapter and the next, I will examine the interaction of Lyell's work with two other historical narratives of the same period, narratives also concerned with the techniques of reconstructing public and private histories: Charles Darwin's account of his voyage on the *Beagle,* and George Eliot's *The Mill on the Floss*. In these chapters we will see the impact of the methodological debate even in cases where it was not immediately and obviously present, where a new way of seeing "the state of things in former ages" promoted itself, like Wordsworthian Romanticism, as both more inductive and more imaginative than its rivals.

## THE GEOLOGICAL CONNECTIONS BETWEEN DARWIN AND LYELL

Darwin was almost immediately influenced by Lyell's theoretical positions in spite of the warnings from his earliest scientific mentors, Cambridge professors John Henslow and Adam Sedgwick, who believed that Lyell's approach brought geology into conflict with Mosaic cosmogony. Henslow gave Darwin the first volume of the *Principles* to take with him on the *Beagle* but urged him "on no account to accept the views therein advocated." Sedgwick, who introduced Darwin to field geology during a geological tour in Wales immediately before Darwin received the offer to join the *Beagle*. fiercely and openly criticized Lyell.[3] Yet Darwin began applying Lyell's principles almost from the moment the *Beagle* touched land. In his *Autobiography* Darwin recalls that his observations in the Cape Verde Islands "convinced" him of "the infinite superiority of Lyell's views."[4] Although this is not strictly true—Darwin's geological notebooks from the voyage indicate that his "conversion" was (appropriately) gradual—a Lyellian coloring is evident in these early observations.[5] By the time he left Chile, however, he could declare himself a thoroughgoing Lyellian, willing even to go beyond his new mentor: "I am become a zealous disciple of Mr. Lyell's views, as known in his admirable book. Geologizing in South America, I am tempted to carry parts to a greater extent even than he does."[6]

As post-*Origin of Species* readers of the *Beagle* narrative, we tend to concentrate on Darwin the biologist, on the observations that served as a foundation for the emergence of his evolutionary beliefs. At the time of the voyage, however, Darwin thought of himself primarily as a geologist, and his letters home indicate this clearly.[7] After his return to England, Darwin became close friends, both personally and professionally, with Lyell. As early as 1838, Lyell saw that Darwin's work on the geology of South America had facilitated the acceptance of uniformitarian principles. At the Geological Society's anniversary meeting in that same year, other geologists were speaking of Lyell and Darwin as the twin champions of uniformitarian ideas.[8] In 1845, when he was thinking of himself more as a natural historian than specifically as a geologist, Darwin dedicated a new edition of the *Journal* to Lyell, "as an acknowledgment that the chief part of whatever scientific merit this Journal and the other works of the author may possess, has been derived from studying the well-known and admirable 'Principles of Geology.' "[9] But this was simply the public acknowledgment of what Darwin had often stated in private. To Lyell himself, Darwin once admitted that his "geological salvation" was "staked" on the fact that Lyell's theories would stand the test of time.[10] To Leonard Horner, a close friend of both men, Darwin wrote: "I always feel as if my books came half out of Lyell's brains & that I never acknowledge this sufficiently, nor do I know how I can, without saying so in so many words—for I have always thought that the great merit of the Principles, was that it altered the whole tone of one's mind & therefore that when seeing a thing never seen by Lyell, one yet saw it partially through his eyes."[11]

## UNIFORMITARIANISM AND GEOLOGICAL DEBATE

What did it mean to see through Lyell's eyes? The central feature of his vision is encapsulated by the full title of the book—*An Attempt to Explain the Former Changes of the Earth's Surface by Reference to Causes Now in Operation*. Uniformitarian (the term was coined by William Whewell in an 1832 review of Lyell's second volume) theory argued that the history of the earth's surface can be accounted for using the same forces we see around us now—erosion, uplift, sedimentation, volcanic phenomena, etc.—operating at similar rates and intensities. Lyell rejected those theories (in Whewell's terminology, "Catastrophist") which contended that geological forces were once more intense and operated over a much wider geographical range than anything we see today, and which used the sudden occurrence of these forces to account for such phenomena as mass extinctions or the formation of mountain chains. In Lyell's view, it was unscientific to theorize about unknown forces when currently visible and measurable forces would do perfectly well. But current forces could explain large-scale phenomena only if they operated over enormous periods of time.

As Martin Rudwick and Stephen Jay Gould have demonstrated, part of Lyell's strategy involved branding all of his opponents as defenders of a six-thousand-year-old earth and searchers for evidence of the most famous global catastrophe, the flood of Noah.[12] By lumping together the antigeologists (who argued that geology was heretical because its teachings openly contradicted Scripture) and the scriptural geologists (who assumed the truth of Scripture as a starting point for geological investigations and then constructed a system of geology in accordance with those truths) with the natural theologians (who argued that, as sources of truth, geology and Scripture must be in accord, but that Scripture, because it does not pretend to be a geological textbook, must be interpreted in light of the truths of science), Lyell angered prominent catastrophist natural theologians like Sedgwick and William Buckland, who had already abandoned their shorter time scales and their efforts to identify specific geological evidence with the Noachian deluge.[13] Thus, while Lyell's accounts of the disputes in early-nineteenth-century geology cannot be relied on, they enjoyed a considerable amount of popular rhetorical success. Among "serious" geologists, Lyell continued to be seen, throughout the 1830s and 1840s, in opposition to the Sedgwicks and Bucklands, defenders of catastrophes and reconcilers, in altered form, of geology and the Bible.[14]

In their assumptions of small changes operating over long periods, uniformitarian ideas in geology and evolutionary ideas in biology are often equated. But until the 1860s, some ten years after Darwin first discussed his ideas on "the species question" with him, Lyell refused to accept the evolution of species. Furthermore, he rejected almost any form of progressionism in the organic and inorganic worlds: not only did he dismiss the notion that one species could evolve into others, he denied that fossil evidence shows that life is growing increasingly complex, and he refused to accept the nebular hypothesis that the earth is cooling down from an original molten state. Geology is a fundamentally historical science, but recent historians of science have demonstrated that Lyell's strict uniformitarianism had profound ahistorical implications. In Lyell's scheme, although the earth's surface undergoes constant change, the earth as a whole is in a dynamic steady-state, with competing forces like uplift and erosion canceling each other out from a global perspective. Not only does the history of the earth lack any overall direction, but individual geological events cannot be differentiated and located in a sequence because, in Stephen Jay Gould's words, to a strict uniformitarian, "[t]he earth . . . has always worked (and looked) just about as it does now."[15] Since one period of sedimentation is indistinguishable from another, the events have no uniqueness and the construction of a history becomes virtually impossible.

The challenge for Lyell was to develop a method that enabled him to preserve the historical uniqueness of geological events and to construct geological chronologies while maintaining his steady-state theory. He divided the tertiary strata based on the percentage of fossils of living species present in each layer, using the assumption that more recent layers will contain higher percentages.[16] Beginning with current species and the present state of the earth, he worked backward through time. This statistical method permitted him to declare in the opening sentence of the *Principles* that "Geology is the science which investigates the successive changes that have taken place in the organic and inorganic kingdoms of nature" (1:1). Geological periods can be differentiated and arranged without abandoning a steady-state view and without endorsing the replication of a pattern. Indeed, Lyell was always careful to reject the notion that his views implied that geological events follow some sort of inevitable cycle: in a letter to Whewell, he insisted that "I expressly contrasted my system with that of 'recurring cycles of similar events,'" arguing that his system is characterized by "an endless variety of effects" rather than a fixed pattern.[17]

## GEOLOGY AND METHODOLOGY

When the Geological Society of London was formed in 1807, its self-proclaimed mission was to put an end to the armchair theorizing that had become an obstacle to the development of a sound, inductively established geological system. What geology needed, its members declared, was fieldwork and observation, a massive collection of data from which, in true Baconian fashion, such a theory would emerge.[18]

At the time of the Geological Society's formation, geologists in Europe generally split into two camps, Huttonians (or "plutonists," because they believed the earth's crust had been formed by the injection and cooling of molten rock from below the surface) and Wernerians (or "neptunists," followers of Saxon mineralogist Abraham Gottlob Werner, who believed the crust had been formed primarily by sedimentation and crystallization from a global sea). In 1802, John Playfair published a readable version of his friend James Hutton's geological system, and Playfair's *Illustrations of the Huttonian Theory of the Earth* turned Hutton's *Theory of the Earth* (first published in 1788, expanded and reissued in 1795) into a monument of Baconian science—in Playfair's presentation, Hutton's views were the result of extensive fieldwork and rigorous induction, and Hutton himself was the Newton of geology. The Huttonians tended to be more prevalent in Britain, but British Wernerians responded to Playfair with their own case for neptunist views as the result of strict induction. It was in this context that the Geological Society called for a moratorium on speculation until observation could yield a sufficiently large body of facts on which to generalize.[19] By 1830, however, Lyell clearly believed that the time had come for a new theory—or more appropriately, *the* theory—which would render the old squabbling forever unnecessary and the old factions forever obsolete. And yet the trick for Lyell was to make his own theory, so dependent on the transformation of the geological imagination, appear inductive and Baconian while portraying the work of almost everyone else as fanciful.

As historians of science in the last generation have repeatedly remarked, Lyell presented the *Principles* as a textbook despite the fact that it offered a new and controversial theory of the earth. In Gould's view, Lyell's book is not only a showcase for a new theory but a manifesto calling for the enshrinement of its methodological assumptions as geology's most basic principles; it offers both a new vision of the history of geology and a thorough survey of geology's fundamental forces, reinterpreted in light of this methodology—"a treatise on method" in the widest sense.[20] In his historical survey of geological theory which opens the *Principles*, Lyell criticizes Werner and praises Hutton, crediting the latter for disdaining hypothetical causes in favor of the action of real, natural forces operating over time. Lyell adopts from Playfair the view that Hutton attempted to provide geology with fixed principles as Newton had for astronomy, that Hutton did for time what Newton did for space (1:56–63). Not surprisingly, Lyell sees the uniformity of forces as proof of his own theory's superior inductive quality and hence its special validity, and specifically condemns his opponents, both living and dead (Werner at 1:56, Lamarck at 2:8, and de Beaumont at 3:339), for not following inductive principles. He even argues that the common assumption in all sciences of the uniformity of nature and its laws implicitly sanctions his version of geological theory: "Our estimate, indeed, of the value of all geological evidence, and the interest derived from the investigation of the earth's history, must depend entirely on the degree of confidence which we feel in regard to the permanency of the laws of nature. Their immutable constancy alone can enable us to reason from analogy, by the strict rules of induction, respecting the events of former ages . . ." (1:165).

In a series of articles written for the *Quarterly Review* a few years before the appearance of the *Principles*, Lyell was already trying to have it both ways, to legitimize his own geological vision by the strict rules of induction. The Geological Society, in its emphasis on geology as "an inductive science founded on observation," had done much to "remove the discredit" cast upon it both by "the wild speculations of earlier authors" and by "the vehemence and passion displayed . . . during the controversy respecting the Wernerian and Huttonian hypotheses. . . ."[21] While lamenting that "vague and unsatisfactory speculations," "uncertain observations," and "inconclusive deductions" are "not yet quite out of fashion," Lyell nonetheless announces that he "must entirely disavow the influence of that fashion, now prevalent in this country, of discountenancing almost all geological speculation."[22] The speculation that Lyell wants to permit is of course his own uniformitarian one that present forces can explain past causes, but he also wants to make clear that this speculation is anything but "vague and unsatisfactory": "[The geologist] must study anew the living works of nature, interpreting the phenomena of former ages by rigidly examining and comparing them with the results of existing causes."[23] Initiating the historical comparison that he will later sound in somewhat different form in the *Principles*, Lyell asserts that such rigid examination and comparison will have the same result in the history of the earth as it has already had for "the fate of former historians of remote ages in the history of man," which, Lyell writes, "should serve as a warning against any hasty conclusions [that current forces are puny compared to those of the past];—the philosophy of more enlightened periods has invariably reduced the gigantic stature and god-like attributes of early heroes and demigods to the dimensions and powers of ordinary mortals; while many of the prodigies and marvels of tradition, at first discredited by skepticism, have been finally shown to be in perfect accordance with the present course and constitution of Nature."[24] Uniformitarianism, implicitly marked as the philosophy of a more enlightened period, reveals the present course and constitution of Nature to be both more mundane and more heroic than geologists have thought.

Other geologists tended to accept Lyell's methodology even though in practice almost all of them were uncomfortable with the extremities of Lyell's views. Participants in the broader methodological debate—recognizing immediately the importance of the methodological controversy in geology for their assessment of Baconian induction—followed a parallel path. The empirically minded John Herschel endorsed Lyell's position most fully, arguing in the *Preliminary Discourse* that "Geologists now no longer bewilder their imaginations with wild theories of the formation of the globe from chaos, or its passage through a series of hypothetical transformations. . . ." Instead of invoking "causes purely hypothetical" and "fanciful and arbitrarily assumed hypotheses," geologists now "confine themselves to a careful consideration of causes evidently in action at present."[25] Yet Herschel did not assume that present causes *must* be sufficient for explaining the earth's past changes; rather, the investigation of present causes is the appropriate first step that will then enable us to conclude whether or not past forces were once more violent. This quiet qualification of Lyell's more dogmatic position was made with greater force by Coleridge and Whewell. For Coleridge, uniformitarianism, as a theory built up exclusively from "facts" without the mental initiative of mind, was necessarily incomplete: "Mr. Lyell's system of geology is just half the truth, and no more. He affirms a great deal which is true; and he denies a great deal which is equally true. . . ."[26] Whewell, in his review of the *Principles*, specified this half truth, complaining that it is just as untenable to assume that geological forces have always been constant as it is to assume that they were once more violent, a view repeated both in his *History* and his *Philosophy of the Inductive Sciences.*[27]

Where Herschel sees in Lyell a Baconian emphasis on facts and a Newtonian emphasis on "true causes" infused with just about the right amount of speculation and imagination, Coleridge and Whewell are critical of a speculation too much grounded in the here and now and therefore, somewhat paradoxically but nonetheless accurately, closing off mental initiative prematurely. These reactions suggest why the methodology of uniformitarianism, enormously successful in its effort to fuse realism and romance, to locate itself both within and without the Geological Society's self-proclaimed Baconian enterprise, was so appealing to a generation of realists like Eliot whose roots lay deeply embedded in the methods of Wordsworthian (rather than Coleridgean) Romanticism. This chapter and the next will be concerned with tracing just such a literary genealogy.

## GEOLOGY AS HISTORICAL INTERPRETATION: THE DISCOURSE OF THE EARTH

Like most geologists, Lyell conceived of his discipline as "the investigation of the earth's history," but reading this narrative of the book of nature posed certain difficulties. Without written records, the geologist had only the present state of the earth from which to infer its past; the "monuments" or "memorials" of geological history did not stand out from the current geological landscape as a human monument—for example, a Roman temple—did.[28] To further complicate matters, geological strata had often been disrupted in such a way as to make the determination of their chronological order difficult, and the fossil record—so important for dating—was imperfect. Catastrophism dominated geological theorizing prior to Lyell because it corresponded to the visible evidence: mountain ranges seemed to be the product of violent uplifting forces, gaps in the fossil record seemed to suggest mass extinctions, the presence of fossils in alpine strata seemed to be explicable only on the basis of universal deluges. Studying geology from the uniformitarian perspective therefore demanded more than observing these monuments of the past; it called for the active interpretation and reconstruction of them as well.

An example of Lyell's approach to "monuments" and "memorials" strikes us in the frontispiece to the first volume of the *Principles,* for Lyell makes the reference literal rather than simply metaphorical. When we open the book, we find an etching of ancient ruins with the caption "Present State of the Temple of Serapis at Pozzuoli." What at first seems an incongruous visual introduction to a book about geology, however, later becomes a crucial feature in the geological and methodological argument of the *Principles.*[29] After a discussion in which he argues that geological knowledge helps us know where to look for ruins of previous civilizations, Lyell uses the Temple of Serapis to chart local changes in the levels of land and sea, and to show that it is the land, not the sea, whose level is altering (1:448–59). Although originally built on dry land, the temple was later partially sunk under the sea and then raised again—Lyell can tell this from the effects of barnacle-like animals which once attached themselves to the temple columns when they were partially submerged. From the "Present State" of a human rather than a geological "monument," Lyell is able to reconstruct part of the Italian coastline's uniformitarian geological past.

When we move to the beginning of the text of the *Principles,* Lyell likens the geologist's task to that of the historian, claiming that just as the historian should be familiar with all branches of human knowledge, the geologist should be familiar with all sciences of organic or inorganic nature, from botany to chemistry to mineralogy to zoology (1:2–3).[30] As Lyell notes in the 1838 *Elements of Geology,* geology is "a science which derives its name from the Greek . . . ge, the earth, and . . . logos, a discourse," and this discourse encompasses much more than the minerals and rocks and soils lying on or in the earth's crust.[31] Fossils, says Lyell in the *Principles,* provide an analogy between the ancient and modern state of the earth, revealing that "a considerable part of the ancient memorials of nature were written in a living language" (1:73). The principle of uniformity provides the geologist with "a key to the interpretation of some mystery in the archives of remote ages" (1:165), while those who reject uniformity are "in the situation of novices, who attempt to read a history written in a foreign language, doubting about the meaning of the most ordinary terms" (3:1).

What most of these comments assume, however, is that the study of the present enables us to understand the past, rather than the historian's more common rationale that the study of the past permits us to make sense of the present and even to predict the future. But moving from present to past is for Lyell the more truly scientific, because inductive, approach, at least in geology: "we cautiously proceed in our investigations, from the known to the unknown, and begin by studying the most modern periods of the earth's history, attempting afterwards to decipher the monuments of more ancient changes . . ." (1:160). Yet the historical survey of geological views which opens the *Principles* moves in precisely the opposite direction, from the ancients up to the present, and Lyell's extended comparison between historical and geological study emphasizes on the very first page this same past-to-present movement:

> By these researches into the state of the earth and its inhabitants at former periods, we acquire a more perfect knowledge of its *present* condition, and more comprehensive views concerning the laws *now* governing its animate and inanimate products. When we study history, we obtain a more profound insight into human nature, by instituting a comparison between the present and former states of society. We trace the long series of events which have gradually led to the actual posture of affairs; and by connecting effects with their causes, we are enabled to classify and retain in the memory a multitude of complicated relations. . . . As the present condition of nations is the result of many antecedent changes, some extremely remote and others recent, some gradual, others sudden and violent, so the state of the natural world is the result of a long succession of events, and if we would enlarge our experience of the present economy of nature, we must investigate the effects of her operations in former periods. (1:1)

This comparison, which seems to us, and would have seemed to Lyell's contemporaries, so full of common sense, in fact stands in direct opposition to the uniformitarian principle that the present is the key to the past. Later in the first volume, Lyell remarks that it is because of the "immutable constancy" of the laws of nature that the process works both ways, that the present can throw light on the past and the past throw light on the present (1:165). But Lyell's equation of history and uniformitarian geology is at this point tenuous because circular: present forces take us back to former conditions whose existence is then validated by arguing with the same forces back to the present. The problem exposed by Lyell's connection of history and geology is perhaps inherent in his efforts to accommodate historical succession within a steady-state uniformitarian model of the earth, the tension involved in asserting that past and present are both the same and different.

This tension is nicely encapsulated by the implications of a quote which ends Lyell's historical section of the *Principles* and is taken from German historian Barthold Niebuhr's *History of Rome*: "he who calls what has vanished back again into being, enjoys a bliss like that of creating" (1:74). Niebuhr enjoyed a great vogue in Britain in the 1820s and 1830s, and his historiography has much in common with Lyell's approach to geology.[32] Niebuhr's work was seen (and indeed he presented it) as "scientific" because of its more rigorous evaluation of sources. Like Lyell's vision of the earth's history, Niebuhr's method was based on the notion that nations pass through similar stages of development, though without simple cyclical repetition from nation to nation on the one hand, or an overall linear progression on the other. Thus Niebuhr was able to use as source material the Roman legends, ballads, and stories rejected by eighteenth-century historians as unreliable. In a strategy recalling both Lyell's comparative dating method and his "present is the key to the past" assumption, Niebuhr set aside those stories which had counterparts in other cultures but treated those unique to Rome as evidence of actual historical occurrences. Furthermore, Niebuhr defended his use of these sources by arguing that they appeared unreliable only because they telescoped what were separate events occurring over long periods of time, an argument remarkably similar to Lyell's contention that the earth appears to have undergone catastrophic change only if we adopt restrictive assumptions about its age.

By associating his discourse with Niebuhr's, Lyell establishes the authority of his work, and particularly his survey of the history of geology, in a reflexive way: he claims affinity with a historical method whose own authority is partly dependent on its similarity to the method of the sciences. But while proclaiming his book as the truly inductive theory that geology has been waiting for, Lyell also asserts his possession of the science's imaginative heritage, its "blissful" creative faculty, still fertile but now under control. It is a legacy inherited, tensions and all, by the Darwin whose books "came half out of Lyell's brains." If Lyell is Zeus and Darwin's books at least half Athena-like, that is perhaps nowhere so evident as in Darwin's at least half "creative" work, his "first literary child," the *Journal* of the *Beagle* voyage.[33]

## DARWIN'S UNIFORMITARIAN IMAGINATION

Specific references to Lyell and the *Principles* appear throughout both the first and second editions of the *Journal*. But these are not mere passing references: they inform what are often extensive uniformitarian "readings" of various geological structures—from the South American coastline and the peaks of the Andes to the islands of Tierra del Fuego and the coral atolls of the Indian and South Pacific oceans—and indicate the extent to which Darwin's theoretical/observational orientation was shaped by Lyell's views.[34] He is often quite direct in affirming the new perspective Lyell has given him. Discussing his Lyellian theory of the formation of the plains of Patagonia, for example, Darwin writes that "[a]t first I could only understand [these phenomena] . . . by the supposition of some epoch of extreme violence, and . . . by as many great elevations, the precise action of which I could not however follow out. Guided by the *Principles of Geology,* and having under my view the vast changes going on in this continent, which at the present day seems the great workshop of nature, I came to another . . . conclusion" (*Journal,* 1:160). "Guided" by Lyell, Darwin replaces his immediate catastrophist explanation with a uniformitarian one. Whereas the catastrophist sees "the great workshop of nature" as a place of diminished activity compared with earlier times, the uniformitarian sees around him the same "tools" producing geological activity at similar rates and intensities. What follows this passage is an extended application of uniformitarian theory, an application which extends beyond Patagonia to the whole of South America and, in the course of relating the continent's geology to the past distribution of its flora and fauna, to the entire world.

Lyell's influence is also evident where it is unacknowledged, for Darwin goes well beyond merely mechanical applications or espousals of uniformitarian principles. Over and over again when describing a new landscape or discussing its current geology, Darwin slips seamlessly into a description of the very different features the landscape wore ages and ages ago, moving in good uniformitarian fashion from present to past. A cryptic example, the slightness of which suggests the pervasiveness of Darwin's uniformitarianism, occurs in Patagonia, when he encounters a belt of sand dunes in the middle of the plains: "The belt of sand-dunes is about eight miles wide; at some former period, it probably formed the margin of a grand estuary, where the Colorado now flows" (*Journal,* 1:68). It is done so quietly that we barely notice it, but from the observation of a group of sand dunes in the middle of a plain Darwin conjures a geological history, with a past landscape very different from that of the present.

A more extended example of Darwin's uniformitarian imagination occurs during an expedition in the Andes. After describing the scenery around him, Darwin writes:

> It required little geological practice to interpret the marvellous story, which this scene at once unfolded; though I confess I was at first so much astonished that I could scarcely believe the plainest evidence of it. I saw the spot where a cluster of fine trees had once waved their branches on the shores of the Atlantic, when that ocean (now driven back 700 miles) approached the base of the Andes. I saw that they had sprung from a volcanic soil which had been raised above the level of the sea, and that this dry land, with its upright trees, had subsequently been let down to the depths of the ocean. There it was covered by sedimentary matter, and this again by enormous streams of submarine lava . . . ; and these deluges of melted stone and aqueous deposits had been five times spread out alternately. . . . I now beheld the bed of that sea forming a chain of mountains more than seven thousand feet in altitude. Nor had those antagonist forces been dormant . . . : the great piles of strata had been intersected by many wide valleys; and the trees now changed into silex were exposed projecting from the volcanic soil now changed into rock, whence formerly in a green and budding state they had raised their lofty heads. (*Journal*, 2:319)

Darwin moves rapidly from present to ancient landscape, then works his way forward to the present again, filling in the necessary geological processes to account for the composition of the current scene. And what at first appears "marvelous" and "astonishing"—in need of some catastrophic interpretation—can in fact be explained, with only a "little geological practice," as a series of gradual processes, processes that are themselves gradually "unfolded" before the eyes of both Darwin and his readers.

Darwin's cinematic recreation of South American geological history only begins to suggest his facility at reconstructing spatially or temporally distant landscapes. On another Andes excursion, Darwin was high enough to observe a fog bank fill the lowlands and valleys, leaving visible only the occasional hillock and the mountains on which he stood. Reminded of the inlets and bays of Tierra del Fuego, Darwin considers the interconnected north-south basins and the transverse valleys which link them to the coast, and argues that Chile must have once resembled Tierra del Fuego when these basins and valleys were filled with water rather than fog. Though triggered by an enormous imaginative leap, this observation is grounded in the existing evidence of Chilean and Fuegian geography and the operation of uniform forces. As Lyell noted in the *Principles*, mankind is at an observational disadvantage because we inhabit only the land—if we could stand at the bottom of the sea in one of Tierra del Fuego's inlets, Darwin argues, the similarity between Tierra del Fuego and Chile would be immediately apparent, and we would not stand in need of imaginative cartography.

On another occasion Darwin relies, in a manner that recalls Niebuhr's use of legends and ballads, on the stories of the local inhabitants to account for phenomena observable in other regions. During his travels through the pampas, he hears about the effects of the severe droughts which periodically strike the region. When water is sufficiently scarce, he is told, animals will rush by the thousands into the few rivers, where, weak from hunger, they become trapped in the muddy banks and drown. This, Darwin claims, explains the presence of vast numbers of bones embedded in the estuary deposits of what were once ancient rivers: the drowned animals eventually float downstream, where they sink to the bottom and fossilize in the sediment. It is the description of the great drought in the pampas during the late 1820s that provides him with the clue that enables him to account for a phenomenon that otherwise seems to demand a catastrophist invocation of deluges. Would not a geologist, Darwin asks, "viewing such an enormous collection of bones, of all kinds of animals and of all ages, thus embedded in one thick earthy mass . . . attribute it to a flood having swept over the surface of the land, rather than to the common order of things?" (*Journal*, 1:123).

But Darwin also becomes on occasion an eyewitness to a "cause now in operation," to the uniformitarian forces that are "the common order of things." During an expedition in Tierra del Fuego, his small exploring party watched as a piece of a glacier fell from a cliff across the bay from their camp, landed in the water with a roar, and sent a large wave traveling towards them. The men were barely able to save the boats before the wave pounded into the beach at their camp. The narrative, though brief, is exciting. Darwin makes clear that the stakes were high, for the party would have been stranded without provisions, surrounded by hostile natives, many miles from the *Beagle*. The published version, however, adds something to this account that the entry in Darwin's *Diary* does not contain: the story has a uniformitarian moral. Darwin says that witnessing this event has enabled him to explain something that had been puzzling him for some time, namely, how the large boulders scattered on the archipelago's beaches could have been recently moved. Because it is missing from the *Diary* entry, this explanation does not seem to have occurred to him at the time of the event, but well after the fact, after he had meditated on its significance. In the *Journal*, this event quietly but vividly confirms that the methodology of uniformitarianism, apparently so unimaginative and prosaic, can make both geological explanations and literary narratives truly extraordinary.

Another *Diary* entry confirms that Darwin, like Lyell, could employ his uniformitarian imagination outside of narrowly geological situations. Sailing along the Patagonian coast on the first anniversary of the day he first heard about the prospect of joining the *Beagle*, Darwin ruminates about the passage of time in his personal life:

> This day last year I arrived home from N. Wales & first heard of this voyage. During the week it has often struck me how different was my situation and views then to what they are at present: it is amusing to imagine my surprise, if anybody on the mountains of Wales had whispered to me, This day next year you will be beating off the coast of Patagonia. And yet how common and natural an occurrence it now appears to me. Nothing has made so vivid an impression on my mind as those painful days of uncertainty; the clearness with which I recollect the most minute particulars, gives to the period of an year the appearance of far shorter duration. But if I pause & in my mind pass from month to month, the time fully grows proportional to the many things which have happened in it. (*Diary*, 87)

What begins rather unremarkably—if you'd told me a year ago that I'd be here today, I wouldn't have believed it—ends in the conversion of his personal history from an example of catastrophism to an example of uniformitarianism. This year of catastrophic upheaval seems to demand a correspondingly catastrophic interpretation, but by separating the two endpoints and contemplating the intervening events, his life's narrative becomes a gradual one, the time no longer compressed but "fully proportional" to all that has occurred.

Although his uniformitarian imagination had to be cultivated, Darwin seems in these examples to have seen through Lyell's eyes without much difficulty. Where his Lyellian vision does become strained, however, is in his eyewitness account of another cause now in operation: the massive Chilean earthquake of February 1835.

It should first be noted that uniformitarianism does include earthquakes, volcanic eruptions, and floods. In the *Principles*, Lyell uses the language of catastrophism to describe such events—they are "catastrophes," "cataclysms," "disasters"—but he is careful to contextualize them: *his* "catastrophes" are produced by causes now in operation, not hypothetical forces of global scope and unimaginable intensity, and they are "catastrophic" only from the limited human perspective. Looking back at Lyell's comparison of geologist and historian, we note that "the present condition of nations is the result of many antecedent changes, . . . some gradual, others sudden and violent" (1:1). The comparison would seem to demand the admission that antecedent geological changes have also been sudden and violent as well as gradual, but Lyell says only that "the state of the natural world," too, is "the result of a long succession of events." Aware of the danger of making such a statement before he has a chance to define the difference between what "sudden and violent" means to him and what it means to his catastrophist opponents, Lyell simply refuses to complete the comparison. Once the definition has been presented, however, the uniformitarian geologist, though free to acknowledge that such phenomena can have devastating effects on human lives and human property, is implicitly urged to resist the interpretation that such effects are devastating in geological terms. What Darwin discovered in Chile is that such a perspective is not easily maintained.

In both the *Diary* and the *Journal*, Darwin gives a vivid account of the magnitude of the quake's geological effects: it was felt over a wide area and was accompanied by volcanic eruptions, tidal waves, and a permanent uplift of the crust, forces which at some places caused changes that would normally occur in a century of ordinary wear and tear (*Diary*, 255; *Journal*, 2:291). But Darwin's initial reaction stresses the sheer incomprehensibility of the event more in its human terms. On first beholding the rubble of the town of Concepcion, Darwin writes in his *Diary* that the destruction is of such magnitude that "it is difficult to understand how great the damage has been," stressing the "impossibility of imagining [the] former appearance and condition" of the town (*Diary*, 255–56). Such despair of reconstructing the past out of present chaos is surprising in any geologist, but especially a uniformitarian one. Yet when Darwin revised the passage for publication, he spoke not of the "impossibility" of imagining what the town once looked like, but of its being "scarcely possible" (*Journal*, 2:291). The alteration, though slight, is significant, for Darwin shifts his attention in the *Journal* to the geological rather than the human catastrophe, using the earthquake to show how improbable would be the kind of cataclysm necessary to lift the entire Andes range to its present height. Instead, Darwin refers to the elevation he has witnessed, "according to the principles laid down by Mr. Lyell," as part of a series of "small successive elevations" (*Journal*, 2:297). Or, as he puts it in the second edition: "it is hardly possible to doubt that this great elevation [the total elevation of the Chilean coast] has been effected by successive small uprisings, such as that which accompanied or caused the earthquake of this year, and likewise by an insensibly slow rise, which is certainly in progress on some parts of the coast" (312). The force required to raise the Andes "by a single blow" would have caused "the very bowels of the earth [to] have gushed out" (314).[35]

When literary critics discuss Darwin's landscape passages, they tend to separate Darwin's "literary" moments from his scientific ones. James Paradis has argued that from the mid-1830s on, Darwin's view of landscape changes from a "Romantic" one—an immediate, highly personal, take-it-all-in-at-a-glance response to nature—to a more "mental" one, less immediate, less personal, in which nature is described using abstract metaphors like species and natural selection. The *Beagle Journal*, therefore, juxtaposes descriptive landscape passages originally written in the Diary—"aesthetic responses . . . to the South American landscape [that] are the deeply felt manifestations of a sensuous bond between perceiver and the perceived"—with the "professional accounts of an area's natural history" developed later.[36] But as the examples I have quoted indicate, Darwin's "aesthetic responses" and "professional accounts" inform each other far more than Paradis will allow. Indeed, it may be said, contrary to Stanley Edgar Hyman's early contention that the *Journal* has "no controlling metaphor or imaginative design," that the dominant imaginative design of Darwin's book, in its "Romantic" as well as its scientific moments, is uniformitarianism.[37] In the case of his Andean survey of South America's geological history, Darwin's remarks are cast as a vision being revealed before his eyes: "I saw . . . I saw . . . I now beheld. . . ." The vision is not so different from the landscape language of "Tintern Abbey": ". . . again I hear . . . Once again / Do I behold . . . Once again I see. . . ." Like Wordsworth's poem, Darwin's vision, though apparently immediate, is in fact a "reading," an "interpretation" of a marvelous geological "story." Because of his practice in reading other geological stories, Darwin is able to compress the immediate experience and the recollection, but it is nonetheless an experience recollected in tranquility. Although he locates himself within this landscape, both as it appears before him and as it must once have appeared, he is guided by the abstract generalization which is uniformitarianism. His initial astonishment gives way to a reading shaped by uniformitarian principles.

A similar comparison can be made between Wordsworth's experience on Snowdon in *The Prelude* and Darwin's observation of the Andean fogs in the *Journal*.[38] In book 13 of the 1805–6 version of *The Prelude*, Wordsworth finds himself above "a huge sea of mist, / Which, meek and silent, rested at my feet. / A hundred hills their dusky backs upheaved / All over this still ocean; and beyond, / Far, far beyond, the vapours shot themselves, / In headlands, tongues, and promontory shapes." This vision becomes the source of a "meditation" that arises later, "when the scene / Had passed away," a meditation in part on how "higher minds" can "build up greatest things / From least suggestions."

Wordsworth's meditation, in other words, is *about* Wordsworthian poetics, about the ability of the poet to call such scenes to mind, find the significance in them, and shape them into poetry. Darwin's experience of standing on a mountain peak above a rising fog also generates a meditation, and the subject of his meditation is also methodological: it is about the ability of the uniformitarian geologist to reconstruct the earth's history from nothing more than the current landscape and forces now in operation. Building up "greatest things from least suggestions," dwelling on the extraordinariness of the ordinary, applies as much to Darwin the uniformitarian geologist as to Wordsworth the Romantic poet.

Reinforcing this reading is Darwin's revision of another South American experience, an experience that more than a decade after the fact induces him to include the voice of a Romantic poet in the text of the *Journal*. During one of his rambles on the plains of Patagonia, Darwin is struck by the "stillness and desolation" of the landscape. In the *Diary*, this produces a brief meditation: "One reflects how many centuries it has thus been & how many more it will thus remain. Yet in this scene without one bright object, there is a high pleasure that I can neither explain or comprehend" (185). In the first edition of the *Journal* there is a slight change: "One reflected on how many ages the plain had thus lasted, and how many more it was doomed thus to continue. Yet in passing over these scenes, without one bright object near, an ill-defined but strong sense of pleasure is vividly excited" (1:155). In the published version, Darwin the uniformitarian carefully replaces "centuries" with the longer and more indefinite "ages," but this change, like the use of "doomed," clearly makes the passage more poetic as well. And though the "I" is excised from the second sentence, the *Journal* passage retains the sense of personal immediacy: it is Darwin experiencing a powerful but indescribable pleasure from a landscape we would not expect to prompt such a response. For the second edition, however, there is a significant change. Darwin reverses the order of the two sentences and changes "one reflected" to "one asked" in order to close the passage with lines from Shelley's "Mont Blanc": "None can reply—all seems eternal now. / The wilderness has a mysterious tongue, / Which teaches awful doubt" (169).

Shelley seems an odd choice for Darwin's uniformitarian imagination to seize upon, for Shelley's emphasis on immediate emotion rather than emotion recollected in tranquility ("Mont Blanc" being a case in point) is very un-Wordsworthian. Yet "Mont Blanc" is in other ways one of Shelley's most Wordsworthian efforts. As a descriptive-meditative landscape poem, it is heavily influenced by "Tintern Abbey," and the use of the landscape as an emblem of the relationship between nature and the mind of the poet anticipates strikingly Wordsworth's already written but not yet published Snowdon passage. Shelley's emphasis is different here: whereas Wordsworth focuses on the mind's creative activity in shaping the "least suggestions" it receives from nature, Shelley is interested in the way nature can overwhelm the mind, the way "My own, my human mind, . . . passively / Now renders and receives fast influencings" (lines 37–38). Yet, even though the emphasis is different, for Shelley as for Wordsworth there is a reciprocity between the mind's passive absorption of nature and its active manipulation of it. When "The everlasting universe of things / Flows through the mind, . . . / The source of human thought its tribute brings / Of waters,—with a sound but half its own" (lines 1–2, 5–6). If the source of human thought brings a tribute to the everlasting universe of things that is only half its own, it is still half. And the mind that passively renders and receives nonetheless is not exclusively passive; it holds "an unremitting interchange / With the clear universe of things around" (lines 39–40).

Since, in addition, Shelley confronts in Mont Blanc a landscape much like Patagonia's in its stillness and desolation—"Mont Blanc appears,—still, snowy, and serene . . . / A desart peopled by the storms alone, / . . . how hideously / Its shapes are heaped around! rude, bare, and high, / Ghastly, and scarred, and riven" (lines 61–71)—there is much to attract Darwin here. And yet where Darwin stands on the plains that epitomize for him a uniformitarian tale of ages of quiescence punctuated by periods of insensible uplift, Shelley's eyes are directed to the top of the Alps. *His* question to which "none can reply" is also a geological one, but it is about catastrophic rather than uniformitarian forces: "Is this the scene / Where the old Earthquake-dæmon taught her young / Ruin? Were these their toys? or did a sea / Of fire, envelope once this silent snow?" (lines 71–74). Darwin's excision of these lines, while obviously called for by the difference in subject matter, also conveniently renders the passage geologically neutral and able to be put into the service of Darwin's own more uniformitarian meditation.[39]

The specific content of Darwin's meditation, as to a large degree it is in "Mont Blanc" as well, is time. Paradis argues that there is conflict between the Romantic artist and the naturalist in perception of time: the artist "discovers time in the self," developing "a sense of one's own aging amid the permanent forms of nature, which altered on a cyclical, seasonal basis"; the naturalist "seeks time in external nature," where it "was beyond the possibilities of human sensation." The "other Darwin" of the *Journal*—"the geologist-naturalist"—competes against Darwin the Romantic artist: "Evolutionary geology and evolution reversed the Romantic emphasis on the permanence and harmony of landscape."[40] Yet Darwin's question is about the permanence of the Patagonian landscape, and Shelley's is about the different aspect once worn by Mont Blanc. Both men experience time in relation to the self and to the landscape, both find reason for pleasure as well as for a sense of doom. When we see in Darwin's descriptions that he is connecting time both to himself *and* to external nature (in the sense that he links his own moment of observation now with what the same landscape looked like long ago), that he is trying to reconcile the sequential and the cyclical and to bring the immensity of geological time *within* the possibilities of human sensation, we must recognize these not as the separate, competing impulses of disparate traditions, but as a dual inheritance already contained within the rhetoric of Lyellian uniformitarianism. The Romantic and the scientific cannot be so easily separated from Darwin's descriptions because each informs the other.

In a notebook entry from 1838, Darwin wrote: "I a geologist have illdefined notion of land covered with ocean, former animals, slow force cracking surface &c truly poetical. (V. Wordsworth about science being sufficiently habitual to become poetical.)"[41] Darwin's language confirms that the uniformitarian imagination must be cultivated, that the geologist must train himself to envision the "slow force cracking surface." But Darwin does not believe that the cultivation of such uniformitarian notions spoils or undermines the poetic faculty. Indeed, having read the preface to the *Lyrical Ballads*, Darwin contends that it is as these notions become habitual that they become "truly poetical." Uniformitarianism is both realistic and poetic, and Darwin's *Journal* endeavors to fulfill the Wordsworthian manifesto.[42]

## DARWIN AND HUMBOLDT

The other great influence on Darwin during the *Beagle* voyage was Alexander von Humboldt. In his *Autobiography*, Darwin recalled reading Humboldt's *Personal Narrative* with "care and profound interest" during his last year at Cambridge.[43] Following in Humboldt's footsteps by journeying to South America, Darwin often expresses a sense of spiritual kinship to the earlier traveler, especially in the Brazilian rain forests. But as Susan Cannon has argued, Darwin's voyage made him more than just a spiritual heir of "Humboldtian science."[44] Humboldt's name was associated with projects of global observation and mapping similar to that on which the *Beagle* was engaged. Like his predecessor, Darwin combined observations from a variety of different disciplines: geology and paleontology, zoology and botany, history and anthropology. And though he visited South America, Darwin traveled through the southern half of the continent, the part not seen by Humboldt, and then continued all the way around the world, as Humboldt had wanted to do.

Darwin's Humboldtian heritage, however, implicitly conflicted with his self-proclaimed status as a disciple of Lyell. Humboldt and his colleague Leopold von Buch had both been students of Werner, and both had carried their mentor's views into distant parts of the globe. For Lyell, Werner's theories were a sort of geological demon that needed to be exorcised by uniformitarianism. In particular, Lyell saw in Werner two dangerous tendencies: the reliance on catastrophist forces (and especially global floods that could be linked to the Noachian deluge) and the ungrounded extrapolation of the geology of a small region into a full-blown theory of the earth. In the *Principles*, Lyell argued that "to travel is of importance to those who desire to originate just and comprehensive views concerning the structure of our globe, and Werner never travelled to distant countries. He had merely explored a small portion of Germany, and conceived . . . that the whole surface of our planet . . . [was] made after his own province" (1:57). Buch's influential catastrophist theory of mountain building, embraced by Humboldt, was also strongly repudiated by Lyell as a violation of uniformitarian principles (*Principles*, 1:386–87). Since Lyell's traveling credentials could not compare to Buch's or Humboldt's, it was especially important that Darwin, carrying Lyell's books with him around the world, be very careful to distinguish what, exactly, he owed to Humboldt. Or better yet, somehow use his Humboldtian heritage to support Lyell.

Three crucial Humboldtian elements of Darwin's narrative—maps, metaphors, and structure—are employed in this way. I'll begin with maps, because the tensions between Humboldt and Lyell, as well as Darwin's strategy for handling them, can be effectively drawn there.

British geologists were most at home with detailed studies of small, preferably British, regions, perhaps augmented by summer fieldwork on the Continent. Lyell was no exception. The maps contained in the *Principles*, like the descriptions to which they correspond, are of small areas. A map of Europe is a rarity. When discussing the Alps, Lyell employs a cross-sectional diagram rather than a map of the whole chain. Humboldt's maps, on the other hand, like his travels, covered wide areas. The *Personal Narrative* contains a large map of Central America and northern South America, as well as maps of major South American river systems. Because of their size, these maps are not incorporated into the body of the text, as Lyell's usually are.

Another important feature of Humboldtian mapping was the use of isothermal lines. Humboldt developed this technique for making comparisons of the climate, flora, and fauna of different geographical regions and different elevations. By connecting places with the same mean annual temperature, Humboldt could explain the geographical distribution of plants and animals. Although this technique is not directly employed in the maps of the *Personal Narrative,* its spirit is present throughout the work in the numerous tables of temperature and pressure measurements from different locations. Lyell, however, was suspicious of isotherms. In the *Principles,* he notes that Humboldt's isotherms often connect places with very different climates (1:107–8). To Lyell, such a technique obscures more important local differences: to say that two cities have the same mean annual temperature is misleading if one city hovers near that mean temperature year-round while the other fluctuates widely above and below it. Facile comparisons were what had led Werner to use the geology of Saxony as a model for the rest of the earth.

The use of maps in Darwin's *Journal* charts a course between Lyell and Humboldt. The first edition contains few maps, and in this respect it is closer to Humboldt's book than to Lyell's. But Darwin's maps are carefully apportioned. Those in the text are Lyellian representations of small areas. Of the two large foldout maps, one is Lyellian and one Humboldtian. The Lyellian map is a detailed representation of the Keeling Islands, which were crucial to Darwin's uniformitarian explanation of the formation of coral reefs and atolls; the Humboldtian map depicts southern South America, completing the map in Humboldt's text. Both large maps were removed from the second edition (over Darwin's objections), leaving only the Lyellian ones in the text.

This balance is also achieved in the verbal maps drawn in Darwin's prose to clarify certain relationships between Europe and South America. In the first edition, for example, in an effort to convey a sense of the great diversity in South American climate and zoology, Darwin transposes the features of the Southern Hemisphere to their corresponding latitudes in Europe. Imagine, says Darwin, southern Scotland covered with glaciers, parrots in a tropical Denmark, icebergs floating in Lake Geneva at midsummer. In the second edition he uses the same strategy to give an idea of the wide area affected by the Chilean earthquake, with the land violently shaken from the North Sea to the Mediterranean. Yet Darwin invariably draws attention to the inadequacy of maps, to their distortions of space. Although he says that one of the "sources of enjoyment in a long voyage" is the fact that "[t]he map of the world ceases to be a blank," Darwin notes that this enjoyment occurs because of the failure of maps to give a proper notion of the "true dimensions" of immense continents, large islands, and jagged coastlines (*Journal*, 2:475). While sailing across the vast expanse of the Pacific, Darwin claims that maps of that region, drawn on a relatively small scale and crowded with the names of islands, fail to suggest the tiny proportion of land in relation to the sea. The antipodes, looked forward to as "a definite point in our voyage homewards," turn out to be an "airy barrier" which, as with "all such resting-places for the imagination, are like shadows which a man moving onwards cannot catch" (*Journal*, 2:388).

When he is not placing limits on the value of his imaginative Humboldtian cartography, Darwin puts these maps in the service of Lyell's uniformitarian vision. Reasoning from the European example has led some scientists, Darwin complains, to "assume that sudden revolutions in climate, and overwhelming catastrophes," have caused mass extinctions of plant and animal life, when in fact South America proves that thick vegetation can exist far from the equator, and that supposedly tropical species can survive in temperate regions (*Journal*, 1:80). Even when he uses his imaginative map to emphasize the strength and scope of the Chilean earthquake, Darwin turns it into a confirmation of uniformitarian principles: if such a force could raise the land only a few feet, then surely a catastrophist account of the formation of the Andes is impossible. And lest there be any doubt of the geological stakes involved here, consider that Elie de Beaumont, another Continental catastrophist taken to task by Lyell, suggested in the 1820s that the sudden uplift of the Andes could have been the geological event responsible for the Noachian deluge.

A second dilemma for Darwin, obvious but crucial, is how to describe for his English audience the scenery, wildlife, people, and customs of far-off lands, how to communicate effectively the beauties of the tropical rain forests or the strangeness of the Fuegian savage. Darwin's most basic solution is equally obvious: he compares things foreign to things English. Thus the guinea fowl of the Cape Verde Islands are likened to "partridges on a rainy day in September," marshes on the pampas are compared to the Cambridgeshire fens, Chilean Indians resemble "old portraits of James the First," and Tahitian cooking fires utilize stones "about the size of cricket-balls" (*Journal*, 1:5, 103; 2:287, 382).

These similes and metaphors, while obviously effective, carry with them the implication that people and things, not to mention physical geography, are not really so different from place to place. If the marshes of the pampas are truly like the Cambridgeshire fens, then perhaps Werner's leap from Saxony to the entire earth was valid. Thus, as with his maps, Darwin is careful to qualify his comparisons. He explicitly cautions his readers, especially in theoretical passages, against allowing his comparisons to obscure the differences among different parts of the globe—the same warning offered by Lyell about Humboldt's isotherms. The presence of fossil remains of large quadrupeds in arid Argentina does not mean that Argentina must once have been covered by tropical vegetation (like India), for large quadrupeds are currently abundant in scantily vegetated areas of Africa (*Journal*, 1:80). Struck by the thick vegetation far from the equator in Tierra del Fuego, Darwin comments on the unsettling implication of this for the geologist from the Northern Hemisphere: "If a geologist were to find in lat. 39° on the coast of Portugal, a bed containing numerous shells belonging to three species of Oliva, . . . he would probably assert that the climate at the period of their existence must have been tropical; but judging from South America, such an inference might be erroneous" (2d ed., 244). To avoid such erroneous inferences even at less scientific levels, Darwin, throughout his comments on Australia, while consistently making use of English referents to describe this English colony, is fastidious about the differences between them, even down to the number of alehouses along their respective post roads. And when attempting to describe the Brazilian rain forests, Darwin compares the sight to that in a hothouse, but "[w]ho from seeing choice plants in a hothouse can magnify some into the dimensions of forest trees, and crowd others into an entangled jungle?" It is possible to portray elements of the scenery, but "to paint the effect [of the whole] is a hopeless endeavour. . . . Epithet after epithet is found too weak" (*Journal*, 2:462–63).

Where Humboldt's influence appears unqualified, however—indeed, where Lyell's vision seems to have little, if anything, to offer Darwin—is in the structuring of the *Beagle* narrative. Humboldt, too, had wrestled with the difficulties of how to organize a narrative of travels. He was determined to avoid writing a "historical narrative," and set about arranging his facts "not in the order in which they successively presented themselves, but according to the relation they bore to each other."[45] But as he composed a variety of treatises on various aspects of his journey (a process reversed by Darwin, who completed the *Journal* before the volumes on the geology and zoology of the voyage), Humboldt felt the need to combine these elements into one narrative, while recognizing that a traveler like himself with disparate interests could hardly produce a simple, unified itinerary. His solution was first to describe the phenomena in chronological order, and then to "consider them in the whole of their individual relations"—what he called the interruption of "historical narrative" with "simple descriptions."[46]

Darwin's narrative, while not nearly so extensive as Humboldt's, does follow this general pattern: most chapters open with an account of his stay in a given area and conclude with more "descriptive" or "scientific" sections on the area's geology, zoology, botany, etc. The chapters themselves are organized by place and follow a linear movement around the globe. Yet Darwin could produce a narrative in which this linear movement of space and time coincide only by significantly distorting the actual chronology of the voyage. The *Beagle*'s job was intimately connected with space and time, but it was this very connection that necessitated the back-and-forth sailing around the South American coastline as well as the return visit to Brazil on the way home. Its purpose was to carry out a series of chronometrical measurements, and as the word "chronometrical" implies, these measurements involved the determination of longitude, or the distance from the Greenwich meridian, by comparing the difference between local time and the time at Greenwich. Taking and checking these measurements, however, required the *Beagle* to frequently double back, visiting and revisiting various places.

Darwin didn't care for this helter-skelter sailing. He welcomed the opportunity to make his overland expeditions while the *Beagle* hopped along the coast. After crossing the equator for the first time, with the knowledge that he was heading back into winter, Darwin admitted that "I am no ways accustomed to this reversed order of things" (*Diary*, 61). When the *Beagle* changed course to stop again at Brazil on its way home rather than continuing straight to England, Darwin complained to his sister that "[t]his zigzag manner of proceeding is very grievous."[47] Even apart from the example of Humboldt, Darwin wanted to impose order and sequence on this chaotic arrangement when he revised his *Diary* for publication; as early as 1837 he wrote to a friend that his organizing principle for the *Journal* would be position, not time.[48]

True to his intention, Darwin constructed his narrative as a circumnavigation of the globe, without backtrackings or multiple visits. We have, for example, one telescoped account of the two trips through the Straits of Magellan, and this account follows chapters on Patagonia and the Falkland Islands, both of which were visited *after* the first journey in the Straits. And the Falklands themselves were visited twice, the first occasion coming before the exploration of Patagonia.[49] These distortions are ignored or alluded to in passing. The first edition excises the months of waiting to sail from England, the two false starts, and the outward passage, opening nearly a month later in Brazil, while the second edition encapsulates the waiting and the false starts into an opening phrase ("After having been twice driven back by heavy southwestern gales . . ."). The retrospect of the journey, which of course appears at the very end of the published version, was written in Darwin's *Diary* before the ship had even returned to England. His telescopings are likewise marked only by a reference to the "first" or "second" visit to a place described just once, or, for the attentive reader, by inconsistencies in the chronology of the *Journal*'s dates. On one rare occasion when he is explicit about his intention to violate temporal order, Darwin briefly asserts the necessity of such a strategy: "To prevent useless repetitions, I will extract those parts of my journal which refer to the same districts, without always attending to the order in which we visited them" (*Journal*, 1:36).

John Tallmadge has argued convincingly that Darwin's "recasting" of the voyage converts "a mere chronological sequence into a spatial and geographical structure without altering the veracity of the events described." He says that the emphasis in Darwin's account is on progression from the unknown to the known, from ignorance to knowledge, with Darwin becoming the hero of his account.[50] Yet this does not explain why Darwin retains the format of a journal, of a "mere chronological sequence." Nor does it explain why Darwin would take such pains, in Tallmadge's view, to "conceal or at least de-emphasize" the influence of Humboldt, when this progressive structural model seems to owe its existence so heavily to Humboldt. As a way of answering these questions, I want to suggest that Darwin is again endeavoring to retain the power of Humboldt's vision while incorporating Humboldt into a Lyellian framework. To develop this argument, let me first return to the text of the *Journal*.

Listening to a mountain torrent in the Andes, Darwin writes:

> The roar which the Maypo made, as it rushed over the great rounded fragments, was like that of the sea. . . . This rattling noise, night and day, may be heard along the whole course of the torrent. The sound spoke eloquently to the geologist: the thousands and thousands of stones, which, striking against each other, make the one dull uniform sound, are all hurrying in one direction. It is like thinking of time, where the minute that now glides past is irrecoverable.

So is it with these stones; the ocean is their eternity, and each note of that wild music tells of one other step towards their destiny.

It is not possible for the mind to comprehend, except by a slow process, any effect which is produced by a cause repeated so often, that the multiplier itself ceases to convey any . . . definite idea. As often as I have seen beds of mud, sand, and shingle, accumulated to the thickness of many thousand feet, I have felt inclined to exclaim that causes, such as the present rivers and the present beaches, could never have ground down such masses. But, on the other hand, when listening to the rattling noise of these torrents, and calling to mind that whole races of animals have passed away from the surface of the globe, during the period throughout which, night and day, these stones have gone rattling onwards in their course, I have thought to myself, can any mountains, any continent, withstand such waste? (*Journal*, 1:303)

In this passage, Darwin hears the river but imagines the sea; he not only conveys the power of this "cause repeated so often," he likens the "one dull uniform sound" of the stones to the eternally gliding time through which it operates. And even the *comprehension* of this fact—a fact which, though simple, in this short passage connects the entire surface of the globe (mountains and beaches, continents and oceans) with its history and the history of its inhabitants—can occur only through a "slow process" of thought. The imagination is here itself subject to uniformitarian law. And by standing in the mountains and listening to this sound, Darwin is slowly brought to an even stronger conviction that such processes can explain phenomena he has already seen, far away on the plains which dominate the other side of the continent.

The passage also combines the directionless steady-state of the "one dull uniform sound," the cyclical repetition of day and night, with the relentless movement of linear time, of progress, change, and history. Though the sound is uniform, Darwin speaks of the stones "all hurrying in one direction," unable to go upstream any more than time can run backward. These stones have a "destiny" that lies in the "eternity" which is the ocean, but that eternity is not really eternal at all: the tiny fragments will be ground down, turned into sand and then compressed into sedimentary rock, and perhaps someday uplifted to become part of the dry land again. While this process goes on in the inorganic kingdom, whole races of animals in the organic kingdom will appear and become extinct. Seen from a larger perspective, these irrevocable forward movements of historically unique events themselves become part of a cyclical steady-state process. This blend of metaphors—of what Stephen Jay Gould calls time's arrow and time's cycle—is characteristic, as we have seen, of Lyell's historical rhetoric. It appears, though not so neatly, in virtually all of the examples I discussed earlier. It is also present, I believe, in Darwin's depiction of "a day-to-day chronology" as an "idealized circumnavigation."[51]

Like jumbled strata, Darwin's chronology is deceptively linear. The journal format testifies both to the influence of Humboldt's example and to the need to create an organized, readable narrative. But Darwin uses the overarching cyclical movement of the *Beagle*, its idealized circumnavigation, to contextualize Humboldtian linearity. As we have seen, Lyell repudiated the notion that the history of the earth's forces is progressive, a linear movement from a violent, literally earth-shaking past to a relatively tranquil present; however, Lyell also denied that his dynamic steady-state model implied "recurring cycles of similar events." Rather, there is an "endless variety of effects" produced from a stable group of forces. For Darwin, then, the return visit to Brazil on the way home is used both to reaffirm earlier themes—the beauties of the rain forest, the cruelty of slavery—and to assert the authority of the now experienced world traveler and scientist. As he ends his own cycle, Darwin is aware of all that he has gained and all that has changed him in the intervening years, but he is also candid about the drawbacks and disappointments of such a voyage. As he later does in the *Origin of Species*, Darwin endorses progressionism, but he does so very cautiously and carefully—much change there has been, but that change has not all occurred in one direction.

Most important, in traveling around the world Darwin was able to do for Lyell's theories what Buch and Humboldt had done for Werner. By applying uniformitarianism with such success in South America and the South Pacific, Darwin helped to make uniformitarianism a theory of the entire earth. His Humboldtian romance, his own "personal narrative," served the ends of Lyellian realism. To be sure, the structure of his narrative easily lends itself to a catastrophist interpretation, for it telescopes separate events into one account. But Darwin knew that through careful reading of jumbled strata and meditation on intervening events, the time would grow fully proportional and the uniformitarian narrative emerge. His application of Lyell's methods to a narrative at once scientific, literary, and personal inscribes Lyell's way of seeing within all three of those forms of discourse. Achieving this melding of the inductive and the imaginative becomes more difficult for George Eliot, however, in her account of the life's voyage of Maggie Tulliver.

# 4

# The "Wonderful Geological Story"

## Uniformitarianism and *The Mill on the Floss*

By the time George Eliot was writing *The Mill on the Floss,* Charles Lyell was the dean of British geologists and uniformitarianism the dominant geological theory in Britain. It was also, of course, the time of *The Origin of Species,* and Darwin, too, relied in his version of evolutionary theory on the changes generated by gradual processes operating over long periods of time, on the cumulative effects of slight variations. Critics of Eliot have often noted the affinity between her novels and the works of Lyell and Darwin, and *The Mill,* with its lingering account of Maggie Tulliver's childhood and its depiction of the social forces operating in St. Ogg's, is a case in point. Yet the flood that ends the novel seems to undercut—and undercut spectacularly—this otherwise uniformitarian narrative. The flood invites us into the world of Lyell's early critics, those catastrophists and natural theologians who defended the geological possibility of the greatest catastrophe of all, the Noachian deluge, and it has seemed to generations of readers the rather crude and sudden intervention of the novelist's designing hand. Absolving Eliot from Henry James's complaint that there is nothing to prepare us for the flood, Barbara Hardy nonetheless describes it as a "deus ex machina" and "the Providence of the novel."[1] Sally Shuttleworth, in her recent work on Eliot and nineteenth-century science, specifically blends these geological and narratological concerns in similar language: the flood acts a "diluvial wave" that, as it disrupts the narrative, "vindicates catastrophe theory."[2]

I want to suggest, however, that far from undermining uniformitarianism or vindicating catastrophism and natural theology, the flood caps a subtle and complex affirmation of uniformitarianism. If generations of readers and critics have generally failed to see this, it has as much to do with contradictions within the uniformitarian model that Eliot was appropriating as it does with some perceived breakdown of Eliot's artistic integrity. In the process of affirming uniformitarianism, Eliot tries to negotiate *between* Lyell and Darwin, critical of Lyell's failure, in 1859–60, to accept evolutionary theory but suspicious of those who equate evolution and progress. As she "uniformitarianizes" the flood, Eliot offers a reading of Darwin similar to the one Lyell would give a few years later in his public acceptance of natural selection. And Eliot's manipulation of the methodologies of Lyellian geology and Darwinian biology is a vital component of her realist project in *The Mill,* a project that claims for the novel the inheritance of both classical and Wordsworthian poetic modes and that stands as a sort of literary manifestation of the debate over Baconianism. But first I must sketch crucial historical distinctions in the history of science of which literary critics have been insufficiently aware.[3]

## LYELL AND DARWIN: FROM UNIFORMITARIANISM TO EVOLUTION

*The Origin of Species* was published while George Eliot was writing *The Mill,* and within a month of its appearance she and Lewes had read it. Lamenting that the work was poorly written, Eliot nonetheless welcomed Darwin to the fold of those, like herself and Lewes, who were already adherents of "the Doctrine of Development."[4] But one scientist who was not yet in the fold of evolutionary thinkers was Lyell himself, the man whose uniformitarianism had helped to make Darwin's version of the development hypothesis possible.

From the first edition of the *Principles,* Lyell strongly opposed theories of the transmutation of species. His attack on Lamarck surely played a role in Darwin's opposition, initially dogmatic but gradually modified in successive editions of the *Origin,* to the notion of the inheritance of acquired characteristics, and in Darwin's search for a different evolutionary mechanism. According to Lyell, species can vary within narrow limits but are not capable of evolving from one form into another. This view coincided nicely with his wider conception of the earth and its inhabitants as a steady-state system, but Lyell's uniformitarianism took him even further. Most geologists and paleontologists believed that the fossil record proved that life on earth had grown increasingly complex, progressing from microscopic organisms to fishes and reptiles, then to birds and simple mammals and finally to humans. In the hands of natural theologians, such a view confirmed humanity's position as the centerpiece of creation, the divinely ordained ruler of all living things. While Lyell agreed that man had appeared late relative to other species, he denied that there was any such progression in the other forms of life, insisting that the fossil record's incompleteness and the then barely begun investigation of older strata did not justify such a conclusion. For Lyell, representatives of all the different orders have always coexisted on the planet, and thus to speak of the "progression" or "increasing complexity" of organic life was inaccurate and unscientific.

Historians of science have identified several distinct uniformities in Lyell's uniformitarianism. One of these uniformities is "nonprogressionism," the rejection of views of the progression of organic life or the successive diminution of inorganic forces.[5] (It is important here to distinguish between nonprogressionism and anti-evolutionism. In Lyell's case, the two were closely linked. But for many others, including Richard Owen, Britain's leading comparative anatomist and later a bitter opponent of Darwin, it was possible to support the idea that life had grown increasingly complex without agreeing that the lower forms had evolved into higher ones.) The assumption that geological forces operate slowly over long periods of time is referred to as "gradualism," and it is this feature of Lyell's theory that prompts critics like Thomas Pinney to note the "close analogy" between the fictional methodology of Eliot and the scientific methodology of Lyell.[6] A third uniformity, "actualism," is the assumption that the rates and intensities of geological forces have always been the same as the rates and intensities we see around us now, and therefore that past changes can be explained solely with reference to present forces. By way of contrast, catastrophists invoked forces of greater intensity and wider scope than anything seen today—and with correspondingly greater effects.[7] Natural theologians assimilated this view into the argument from design by contending that the Creator had used catastrophic forces to prepare the earth for habitation by human beings, and that once the earth was ready, these forces diminished.

These positions were fully articulated in the *Principles*, but Lyell continued to assert them well into the 1850s, most prominently in his 1851 presidential address to the Geological Society. His doubts increased during the 1850s, as he talked with Darwin and pondered "the species question" himself, but the appearance of the *Origin* marked an important public break between his thought and Darwin's, for Darwin's book, while retaining a strong commitment to Lyell's gradualism, not only abandoned the uniformitarian opposition to evolution, but called into question strict Lyellian notions of nonprogressionism and, to a lesser extent, actualism. It should come as no surprise, then, that Lyell was not an immediate convert to Darwin's views, and that when he did convert, he sought to incorporate natural selection into his own scheme without making too many concessions. Lyell announced his change of mind in *The Antiquity of Man* (1862), two years after the publication of *The Mill*, but it came with a decidedly nonprogressionistic reading of Darwin: in Lyell's words, "one of the principal claims of Mr Darwin's theory to acceptance, is that it enables us to dispense with a law of progression as a necessary accompaniment of variation."[8] Contemporaries like Herbert Spencer, however, recognized the significance of Lyell's careful conversion. Because "the Uniformitarians . . . not only reject the hypothesis of development, but deny that the modern forms of life are higher than the ancient ones," Spencer claimed that "Sir Charles Lyell is no longer to be classed among Uniformitarians. With rare and admirable candour he has . . . yielded to the arguments of Mr. Darwin."[9] Although Darwin's evolutionary views can certainly be read—and often were read—as an endorsement of progressionism, at the time Eliot was at work on *The Mill*, Lyell's strict uniformitarianism stood in opposition to all versions of "the hypothesis of development," including Darwin's.[10] In short, then, we must be very careful when we equate Lyell's uniformitarianism with Darwin's natural selection, especially in 1859–60.

## GENESIS, GEOLOGY, AND GEORGE ELIOT

Eliot's interest in geology began quite early in her life and has a discernible history of its own. As a young woman, she read many books about geology, in particular those which sought to reconcile recent advances in the science with the scriptural accounts of the creation and early history of the earth in Genesis. She read, for example, both Vernon Harcourt's *The Doctrine of the Deluge* (1838) and John Pye Smith's *Relation between the Holy Scripture and Some Parts of Geological Science* (1839) (*Letters*, 1:34, 110). An 1841 letter indicates that she had recently read both William Buckland's Bridgewater Treatise, *Geology and Mineralogy Considered with Reference to Natural Theology*, and Lyell's *Principles* (*Letters*, 8:8). As a young woman Eliot considered herself rather an expert on the controversy, telling a friend that her familiarity with the arguments enabled her to "swallow" Harcourt's book "without much mastication" and to conclude that it "supports or rather shakes a weak position by weak arguments" (*Letters*, 1:34). Gordon Haight is probably correct when he notes, apropos of Smith's book, that during Eliot's movement away from Evangelicalism such a work "encouraged speculation about the very doubts it was meant to dispel."[11]

Buckland's book, which Eliot "read . . . with much pleasure," though it lies in a similar tradition, does not spend much time in raising theological issues and offering extensive Biblical reinterpretations (*Letters*, 8:8). While Smith considers at length how the Creation and the Flood are to be interpreted to bring them into accord with geology, Buckland's treatise skirts both issues. The great length of geological time is simply accepted at the outset, with the statement that Genesis and geology are not at odds because the six days of creation need not be taken literally. The Flood receives only a footnote, in which Buckland recants his earlier view that some geological event can be identified with the Deluge.[12] Instead, Buckland's natural theology concentrates more on the way the Creator adapted every fossil species, just as he has done with every living species, to its environment, and on the way he has adapted the earth to the habitation of man. The earth was once wracked by catastrophic forces, but the Creator used these forces to prepare the earth for man's (and especially the Englishman's) benefit, to form deposits of coal and minerals and to transport topsoil to fertile valleys. In the distant past an "uninhabitable planet," the earth, "through long successions of change and of convulsive movements," has been brought to "a tranquil state of equilibrium, in which it has become the convenient and delightful habitation of man" (1:49–50).

Eliot's pleasure in Buckland's treatise is both telling and understandable. Commissioned to demonstrate "the Power, Wisdom, and Goodness of God as Manifested in the Creation" and written by eminent scientists like Buckland and William Whewell, the eight Bridgewater Treatises, by extending the arguments of natural theology with up-to-date examples from throughout the sciences, represented the final high point for the design argument before the deathblow dealt by the *Origin*. Buckland's was the last of the eight to appear and also the most eagerly awaited; the first printing of five thousand copies was sold out before publication. Yet in conservative religious circles these attempts to synthesize science and religion were considered scandalous simply because they took geology rather than Genesis as their starting point, so Eliot's early enthusiasm for Buckland simultaneously reflects a desire to be scientifically and philosophically sophisticated while retaining a sense of the presence of moral purpose in the natural world.[13]

The difference between Buckland's catastrophist natural theology and Lyell's uniformitarianism was not lost on Eliot.[14] Comparing Lyell's book to Buckland's, she says that Lyell's "is good though it differs in Theory" (*Letters*, 8:8). What makes this comment particularly striking is that such a distinction is not made explicitly in either man's book: Lyell consciously avoided specific criticisms of his British colleagues;

Buckland makes such extensive use of Lyell that it is not immediately obvious that the two men are in different theoretical camps.[15] The tone of the comparison, however, seems to suggest some degree of preference for Buckland's natural theology, a preference still evident ten years later, when, in an 1851 review article, Eliot echoes Buckland's language, saying that a "correct generalization gives significance to the smallest detail, just as the great inductions of geology demonstrate in every pebble the working of laws by which the earth has become adapted for the habitation of man" (*Essays*, 31). This is the passage that causes Pinney to remark that Eliot was here thinking of Lyell's *Principles* and to mention the close analogy between their respective methods, and there is in the pebble an echo of Lyell's comment that "every pebble tells its own tale" of the small forces operating through long stretches of geological time. But the notion that a pebble can provide a narrative of the earth's history in miniature was common in geology and could just as easily be incorporated into arguments from design, where the "smallest details" of the Creation point to the existence of an omniscient and beneficent Creator. In his Bridgewater Treatise, Buckland points out that in Paley's famous watch argument, the contrast between a stone and a watch is unnecessary because the stone itself provides direct evidence of design: "if the stone were a pebble, the adventures of this pebble may have been many and various, and fraught with records of physical events, that produced important changes upon the surface of our planet" (1:572). Lyell's pebble tells a uniformitarian tale; Buckland's has many possible narratives—some of which he details—but they all provide evidence of design. That Eliot has Buckland in mind here, however, is evident from the second part of her statement, from the notion that geological laws have worked for the *benefit* of man, that the earth has been adapted for *man's* habitation.

It is shortly after this, however, that Eliot begins to move away from the geological traditions of Buckland's catastrophist natural theology and of scriptural geology, and, with qualifications, toward Lyell's uniformitarianism. In 1852 Edward Forbes wrote an article on geology for the *Westminster Review,* an article which reviews several recent geological works, among them Lyell's *Manual of Elementary Geology* and the eighth edition of the *Principles*. Eliot commended Forbes's article highly both before and after its publication (*Letters*, 1:370, 2:34). In his review, Forbes surveys the history of modern geology, discussing briefly the work of the science's giants, including Buckland, Sedgwick, and Lyell. Although he avoids references to the controversies and disputes among these men, Forbes's prejudices are evident: Buckland and Sedgwick are lauded for the style of their public presentations, but Lyell is singled out both for his practical geology (his work on the tertiary strata, which, though comparatively noncontroversial, was nonetheless inseparable from, and necessitated by, his uniformitarian approach) and for his theoretical work (his "admirable writings . . . have done more to diffuse a sound spirit of geological inquiry and reasoning than any ever published").[16] Forbes goes on to criticize those who object to geology on religious grounds as well as those geologists who try to make their research agree with Scripture, singling out John Pye Smith as the worst example of the latter group. He echoes Hutton's famous dictum—adopted by Lyell for the frontispiece to the *Principles*—that the earth reveals "no vestige of a beginning, no prospect of an end," when he says that "there is sign neither of beginning nor of end to be discerned in purely geological phenomena."[17] He adopts Lyell's criticisms of Lamarck's evolutionary hypothesis, shares Lyell's (public) view that species were separately created, and questions, with Lyell as his justification, the belief that there has been progression in the organization of plants and animals through time, excepting—as Lyell does—man.

Following the appearance of Forbes's article, Eliot's criticism of reconciliations of Genesis and geology becomes both open and harsh in her reviews of "Rev. Cummings' Evangelicalism" (for the *Westminster* in October 1855) and James Heywood's *Introduction to the Book of Genesis* (in a January 1856 issue of the *Leader*). Sounding in the article on Cummings very much like Forbes, she condemns those who object to geology on religious grounds as well as those who try to develop a scriptural geology. Discussing Cummings' *The Church before the Flood*, which she describes as being "devoted chiefly to the adjustment of the question between the Bible and Geology," Eliot criticizes Cummings' mode of treatment rather than directly assessing the controversial question itself. According to Eliot, Cummings attempts to reconcile Genesis with the discoveries of science by means of "imaginary hypotheses and feats of 'interpretation'" (*Essays*, 176). An indication of his "confused notions" is his insistence that there "is not the least dissonance between God's written book and the most mature discoveries of geological science," although there "may be a contradiction between the discoveries of geology and our preconceived interpretations of the Bible" (*Essays*, 176–77). This willingness to alter such "preconceived interpretations" is called by Eliot in her review of Heywood's book the "'accommodation' theory." It assumes that biblical revelation was adapted to the understandings of the people to whom it was originally revealed, and that therefore it must be updated and revised to keep it in accord with modern knowledge. This system of interpretation, which Eliot specifically identifies as the method of Smith but is also applicable to Buckland, is "very elastic" and "necessarily leads to what may be called a mitigated orthodoxy or a mild heterodoxy" (*Essays*, 256–57). While she approves of accepting scientific truths and of opening Scripture to historical interpretation, Eliot is clearly hostile to elaborate reconciliations, especially when they are, like Cummings', inconsistent and illogical.

As Eliot put distance between herself and the geological positions of Smith and Buckland, she also moved, with important qualifications, toward Lyell. Much of the evidence from the 1850s relates to Lewes, whose scientific positions were generally shared by Eliot.[18] Lewes was an enthusiastic proponent of uniformitarianism, endorsing it both in his own scientific projects and in his theoretical discussions of scientific method. In the *Leader* of 13 December 1856 Lewes wrote a highly laudatory review of David Page's *Advanced Textbook of Geology*, which he says deserves "the very first place among geological works addressed to students." Page's work is unquestionably uniformitarian in its assumptions, and a passage excerpted by Lewes underscores the uniformitarian approach to the construction of historical narratives so emphasized by Lyell: "Without this uniformity in the great operations of nature, the history of the PAST would be uncertainty and delusion. We can only read the past as connected with the present; and premise of the future from what is now going on around us."[19] A decade later, Lewes' endorsement of actualism remained strong. In the prolegomena to his *History of Philosophy from Thales to Comte* he remarks that "[t]o read the pages of the great Stone-book, and to perceive from the wet streets that rain has recently fallen, are the same intellectual processes. . . . [T]he mind . . . infers that the phenomena were produced by causes similar to those which have produced similar phenomena within recent experience. . . . The geologist . . . proceeds on the assumption that the action of waters was essentially the same millions of years ago as it is in the present day; so that whatever can be positively proved of it *now*, may be confidently asserted of it *then*."[20] In a passage from *Sea-Side Studies*, which first appeared in *Blackwood's* in 1856, Lewes' uniformitarianism is again evident, though on this occasion he emphasizes gradualism rather than actualism. Discussing the ability of certain molluscs to bore into wood and rock, he notes that "zoologists have allowed themselves to be thrown off their balance by contemplating the stupendous results produced by creatures so insignificant. But after learning the history of the formation of coral reefs and islands, we begin to appreciate the influence of minute agencies continued through long periods of time." The reference to coral reefs and islands already links the gradualism of geology to that of biology (and also brings in Darwin, whose coral reef theory, developed on the *Beagle* voyage, was a powerful application of uniformitarian methods), but Lewes goes on to make the connection even more explicit: "just as falling water wears away granite, by the incessant repetition of gentle blows, so do these molluscs excavate rock or wood by the incessant repetition of muscular friction."[21]

Where Eliot and Lewes differed from Lyell during this period was on the subject of the "doctrine of development." In the *Leader* of 7 September 1850, Lewes reviewed the fifth edition of Adam Sedgwick's *A Discourse on the Studies of the University of Cambridge*. Sedgwick used the occasion of the new edition to write a vastly expanded introduction, in which he repeated his earlier attack on the evolutionary ideas in Robert Chambers' *The Vestiges of the Natural History of Creation*—an attack so forceful that Darwin was careful to publish his own evolutionary theory only when he was sure he could meet all of the objections enumerated by Sedgwick. Lewes' review focuses almost exclusively on Sedgwick's attack, defending the *Vestiges* by counterattacking Sedgwick.[22] A year later, on 18 October 1851, another *Leader* review also defends the *Vestiges*. This article—written, if not by Lewes' hand, during his editorship—reviews both Lyell's 1851 presidential address to the Geological Society and an evaluation of that address in the *Quarterly Review* by Richard Owen. Owen applauds Lyell's use of uniformitarian principles in the inorganic world but sharply criticizes his adherence to nonprogressionism in the organic world, launching his criticism with a quote from Sedgwick's *Discourse*. The *Leader* review, in turn, uses the occasion to criticize *both* Lyell and Sedgwick for opposing the *Vestiges*, though it singles out Lyell's position as the more dogmatic. "[O]ur purpose," writes the reviewer, "is to draw attention to the triumphant demolition of Sir Charles Lyell's attack upon the doctrine of a gradual development in the scale of being, both animal and vegetable, from the earliest periods to our own time." Owen's progressionism is at least potentially compatible with evolutionary schemes, but Lyell's "obstinate persistence in his objection to the Development Theory" is not.[23]

Lyell's nonprogressionism was also a source of disagreement, particularly for Lewes. In *Comte's Philosophy of the Sciences* (1853), Lewes asserts the evidence for progressionism in biology: "Simplicity is not the principal measure of real perfection; biology studies show, on the contrary, that the increasing perfection of the animal organism consists in the increasing speciality of the various functions accompanied by organs more and more distinct."[24] Yet the evolutionary thought of Comte and Spencer obviously embraced much more than biology and geology, and equated evolution not only with progress, but often with necessary progress. Darwin, while breaking from Lyell's dogmatic non-progressionism, was unwilling to go to the opposite extreme, and Eliot, herself inclined to less optimistic views about progress in either the organic kingdom or human society, immediately picked up on this. "Natural Selection is not always good," she once wrote, "and depends (see Darwin) on many caprices of very foolish animals" (*Letters*, 4:377).

That Eliot did indeed have thirty years of geological and biological debate on her mind at the time she was writing *The Mill* is evident from her notebooks, which indicate that she was reading Mary Somerville's *Physical Geography* during this period.[25] Somerville's widely read popularization includes an account of the last half-century's geology which draws eclectically on the works of the present and preceding generations' best-known geologists. More important, her account has affinities with Eliot's own evolving position on geology. Although the book's overall flavor is uniformitarian, and its largest debt is due to the geological works of Lyell and Darwin, Somerville often juxtaposes the language of uniformitarianism with that of catastrophism and natural theology—a situation reminiscent of Eliot's early enjoyment of both Lyell and Buckland. On the one hand, for example, earthquakes, floods, and volcanoes—"the august and terrible ministers of Almighty Power"—have "opened the seals of the most ancient records of creation, written in indelible characters on the 'perpetual hills and the everlasting mountains.'" Such current forces are only small versions of the "violent convulsions" of the distant past, "the tremendous forces that must have been in action at epochs immeasurably anterior to the existence of man" but which have since diminished to bring "the rude mass to its present fair state." On the other hand, however, "these and all other changes that have taken place on the earth have been gradual and partial, whether brought about by fire or water. . . . [T]hese convulsions have never extended all over the earth at the same time—they have always been local. . . . There is no proof that any mountain-chain has ever been raised at once; on the contrary, the elevation has always been produced by a long-continued and reiterated succession of internal convulsions with intervals of repose."[26] Somerville's text usually does not mark such contradictions, nor does it explicitly offer a sense of the historical development of these views, but Eliot, with her extensive exposure to them, would have had no trouble reading between the lines.

GEOLOGY AND *THE MILL ON THE FLOSS*: HISTORICAL VISIONS

Lyell denied both that the earth's history was directional and that it was cyclical. The history of his steady-state earth consisted of unique geological events which could be differentiated and dated, but which did not provide evidence of progress. The historical vision of *The Mill* is similarly constructed in one of the novel's many geological references, an account of Aunt Glegg's clothing: "One would need to be learned in the fashions of those times to know how far in the rear of them Mrs. Glegg's slate-coloured silk gown must have been; but from certain constellations of small yellow spots upon it, and a mouldy odour about it suggestive of a damp clothes-chest, it was probable that it belonged to a stratum of garments just old enough to have come recently into wear."[27] The geological joke at the expense of Mrs. Glegg's wardrobe continues a few pages later, when we learn that Mrs. Tulliver is "too proud to dress her child in the good clothing her sister Glegg gave her from the primeval strata of her wardrobe" (1, 7).

Lyell's uniformitarianism made the present the key to the past. The narrator, by opening with the phrase "one would need to be learned in the fashions of those times," seems to suggest that expert knowledge of the past is necessary to determine just how out of fashion Mrs. Glegg's gown is. But as the narrator continues, we see that she actually depends on the data of the present—the yellow spots and moldy odor—to locate Mrs. Glegg's gown in a historical sequence. The "strata" of Mrs. Glegg's clothes chest, laid down when the gowns were new, are gradually eroded over time (literally "worn" down), exposing new gowns that are increasingly out of date. These fossilized relics of the past suggest change without improvement, cycles without repetition. This was precisely Lyell's view of the history of the earth and its inhabitants.

Like most of the other references to geology and geological texts in the novel, this pleasantry is part of the narrator's effort to position her narrative in a specific historical past. Comments on clothing styles are a common method in *The Mill* for emphasizing the disparity between the present of the narrative and the present of the narrator and reader. Mrs. Tulliver, for example, is initially described as "a blond comely woman in a fan-shaped cap," after which the narrator remarks that "I am afraid to think how long it is since fan-shaped caps were worn—they must be so near coming in again. At that time, when Mrs. Tulliver was nearly forty, they were new at St Ogg's, and considered sweet things" (1, 2). Mrs. Pullet, we are told, is "a striking example of the complexity introduced into the emotions by a high state of civilisation—the sight of a fashionably drest female in grief. From the sorrow of a Hottentot to that of a woman in large buckram sleeves, with several bracelets on each arm, an architectural bonnet, and delicate ribbon-strings—what a long series of gradations!" (1, 7). Yet the "complexity" of this "high state of civilisation" is immediately undercut, for Mrs. Pullet's superior development consists merely in her ability to clear a doorway in a state of grief without crushing her buckram sleeves, though the narrator does explain parenthetically that "at that period a woman was truly ridiculous to an instructed eye if she did not measure a yard and a half across the shoulders." Although they are somewhat awkward and intrusive, these comments, by combining geological metaphor and realistic detail, keep before the reader the historical otherness of the narrative as well as the narrator's position as privileged interpreter and reconstructor of the past.

The combination of the revolving mill wheel and the flowing river also suggest *The Mill*'s complicated historical vision and the narrator's role in interpreting and communicating that vision. As with Mrs. Glegg's clothes chest, there is a fusion of the cyclical and the sequential. The river always flows; the wheel always turns. The mill itself is incorporated into such a pattern: there is the cycle of repetitions, the holding of the mill in Tulliver hands through generations, and the sequential movement of history, a historical succession marked by floods, changing owners, and lawsuits. Changes in the ownership of the mill are said to make the river angry and bring the floods, and this has indeed been the case: the mill is in fact not one mill but several—Mr. Tulliver recalls the "old half-timbered mill" pulled down by his grandfather after "the last great floods" (3, 9), and the new mill of four generations must itself be rebuilt after the flood that kills Maggie and Tom. Generational continuity provides a frame within which change can occur, as when Tom inscribes his father's curse against Wakem in the genealogical record of the family Bible. The river is more closely identified with the irreversible sequence of time—Maggie's "destiny," the narrator tells us, is like the course of an "unmapped river," sharing with all other rivers a "final home" in the sea (6, 6)—but not exclusively so. The language describing Maggie's destiny recalls Darwin's Andean torrent, in which small stones rattle "towards their destiny," to "the ocean [that] is their eternity," yet which Darwin also imaginatively transforms into an emblem of uniformitarian change, able to wear down mountains and carry them to the sea, an epitome of the irreversibility of time contained within cyclical, steady-state geological processes. The Floss, with its periodical floodings and tidal fluctuations, plays a role in connecting the linear sequence of human lives with the large rhythms of natural history.

As in her other novels, Eliot carefully maps out historical differences because such differences are important—we cannot understand the individual lives within the social organism if we do not understand the external forces which make the social organism what it is and which impinge upon the individual. But her historical vision depends, as does the power of her fiction, on an essential sameness in the human lot. In *The Mill*, this position is gradually established by the narrator's daydream, which not only superimposes two February days, one of the present, one from the historical past, but also deliberately conceals the temporal differences between them by using the present tense to describe both settings. So, too, the narrator suggests an affinity between herself and the little girl, whom we will soon come to know as Maggie: both stare in rapt attention at the revolving mill wheel, but the narrator's view from the stone bridge, which is itself contained in her insistently personal memory ("I remember those large dipping willows. I remember the stone bridge . . ."), takes in both Maggie and the wheel.[28] When the narrator reveals, however, that she is leaning not on the stone bridge but against the arms of her chair, that she is not observing but recalling the scene of the girl and the mill wheel, we are perhaps led to speculate that the narrator is in fact the grown-up girl. In one sense, this is true: the novel's autobiographical elements connect Maggie to the narrator and the narrator to Marian Evans. But the narrative will eventually deny the strict identification of Maggie and the narrator, for Maggie is dead. We are dependent on the narrator's separation of herself and Maggie just as we are dependent on her to separate those two February days.

The process of separating what appear to be two adjacent historical events, and filling in the narrative of events that occur between them, is vital, as we saw in the previous chapter, to the transformation of a catastrophist narrative into a uniformitarian one that is simultaneously realistic and romantic, extraordinary in its own way. In Eliot's novel, the legend of Ogg offers a catastrophist vision of history in which floods are an everyday occurrence, but the "air of romance" disappears when these floods are reconstructed into a sequence of similar events separated by years of gradualist change. We recall Niebuhr's project of constructing a history of early Rome out of the catastrophic, telescoped events contained in ballads and legends, as well as Darwin's rumination in his *Diary* about those two August days, the day he first heard of the *Beagle* and the same day one year later when he stood aboard it, sailing along the South American coast. It is only through meditation, through the construction of a narrative of intervening events, that the time, in Darwin's words, "fully grows proportional to the many things which have happened in it." In Eliot's novel we are similarly dependent on the narrator to separate the two February days and to detach the little girl from herself, by filling up part of the thirty-year blank with a gradualist narrative of "the many things" that have occurred during it.

## GRADUALISM

One aspect of uniformitarianism that Eliot's fiction clearly does embody, both in subject matter and in structure, is gradualism, the operation of slow processes over long periods of time. Her focus is on the trivial and the small, the "hidden" tragedy of those "insignificant people, whom you pass on the road everyday," rather than "that conspicuous, far-echoing tragedy, which sweeps the stage in regal robes" (3, 1). The narrator likens her approach to the methodology of science, for science shows that the small, because it is part of a matrix of relations that includes the great, is also important. Although this attention to trivial things is, the narrator tells us, antithetical to the way our imagination usually works, it is the smallest things that have such effect in human lives and which, by implication, are essential to the action of this novel. Maggie's secret meetings with Philip in the Red Deeps are uncovered quite indirectly and quite unconsciously by Aunt Pullet, not, as Maggie fears, by a catastrophic confrontation with Tom or Mr. Tulliver, thus proving that "[t]hose slight indirect suggestions which are dependent on apparently trivial coincidences and incalculable states of mind, are the favourite machinery of Fact, but are not the stuff in which the imagination is apt to work" (5, 5).

This emphasis on the effects of "slight indirect suggestions" and "trivial coincidences" recalls the gradualism both of Lyell's geology, where the trickling stream can eventually level mountains, and of Darwin's natural selection, where small variations can eventually transform a species. Eliot's realist aesthetic—"the machinery of Fact"—seems to diminish the imagination, like Darwin's mill-like description of his mind as a machine for grinding general laws out of large collections of facts. But the achievement of the uniformitarian imagination lies in showing that the "grinding," trivial, less obvious forces around us produce the same effects we usually ascribe to sudden and violent upheavals. For Eliot as for Lyell and Darwin, a gradualist vision is both more "real" and more imaginative because it sees beyond or beneath that which our limited minds label as catastrophic. The machinery of fact is in a sense the aesthetic corollary of the Baconianism, simultaneously inductive and imaginative, whose construction we have traced in Chapters 1 and 2. And it is no accident that Eliot's Wordsworthian interest in the extraordinariness of the ordinary, in the romance of realism, occurs in a novel so deeply suffused with the language of Lyellian geology and Darwinian biology.

Eliot repeatedly unfolds apparently catastrophic events as the product of gradual movement. Even such events as the family quarrel, Mr. Tulliver's bankruptcy and his attack on Wakem, Maggie's departure with Stephen, and the flood itself are, when reconstructed through the narrator's uniformitarian perspective, products of gradualist processes. We are prepared for these "catastrophes" both by prior events (Maggie and Stephen are left alone, for example, because of Philip's jealousy and Lucy's efforts to play matchmaker between Maggie and Philip) and by patterns of imagery (as with the flood). The care taken by Eliot to structure her narrative as a gradualist one, and the difficulties critics have in separating an awareness of that care from their own sense of the novel's "catastrophes," can be seen in a comment by W. J. Harvey about what he calls "Tulliver's catastrophic bankruptcy." Writing of the second family meeting, Harvey notes that it has its roots in the first family meeting, where Mr. Tulliver's quarrel with Aunt Glegg sets in motion the series of money problems which will culminate in his bankruptcy.[29] In the effects it has on the Tulliver family, the bankruptcy is indeed a catastrophe, but in terms of narrative structure, it is clearly *not* a catastrophe. The structural symmetry of the two scenes is causally linked to the slowly unfolding succession of narrative events.

The term "gradualism" fails, however, to capture the diversity in rate of change allowed to geological forces in Lyell's and Darwin's uniformitarianism. "Gradual" implies not only slow movement, but also constant movement at a basically constant rate. Yet for Lyell, geological processes can diminish, accelerate, or even stop entirely; conflicting processes (for example, uplift and erosion) can occur at the same time and in the same place. Writing in the *Principles*, Lyell says that "[w]e learn from the study of mutations now in progress, that one part of the earth's surface may, for an indefinite period, be the scene of continued change, while another, in the immediate vicinity, remains stationary. We need go no farther than our own country to illustrate this principle. . . ."[30] Compared with Dickens, whose novels depend on narrative "catastrophes"—surprises, suddenly revealed secrets and connections, providential turnings of fortune, coincidences that are never trivial—Eliot appears to rely on a gradualism almost monotonous in its gradualness. Yet gradualism offers diverse possibilities for the movement of human as well as geological narrative, and, like Lyell, Eliot did not need to go any farther than her own country to illustrate this principle. Eliot usually exploits these possibilities in *The Mill* by way of the characters' differing senses of the passage of historical time in their own lives. At one extreme is Mr. Tulliver, whose life is one of painfully slow movement. In conversation, he is "not a man to make an abrupt transition" (1, 3); even his death "was not to be a leap" but "a long descent under thickening shadows" (3, 4). Although bewildered by a world which he sees as puzzling in its complexity, Mr. Tulliver endeavors to think generationally—if he cannot keep up with the changes going on around him, or if he cannot beat Wakem, Tom will do so some day. In his pessimistic vision, "the country could never again be what it used to be" (1, 7), and although he finds Mr. Deane's contrary assessment soothing, its denial of the irreversibility of social change merely delays the inevitable social decline.

In contrast to Mr. Tulliver, Mr. Deane sees the world about him not only progressing, but accelerating in its progress. His perspective, however, is primarily economic. "[T]he world," he tells Tom, "goes on at a smarter pace now than it did when I was a young fellow. Why, sir, forty years ago . . . [t]he looms went slowish, and fashions didn't alter quite so fast. I'd a best suit that lasted me six years. Everything was on a lower scale, sir—in point of expenditure, I mean. It's this steam, you see, that has made the difference: it drives on every wheel double pace, and the wheel of fortune along with 'em . . ." (6, 5). The movement of time is again marked by changing fashions, though on this occasion through the intermediary of steam, which, in driving the looms faster, accelerates the cycles of fashion, with consequences for Aunt Glegg's clothes chest. Wheels revolve with greater velocity, but the spinning wheels translate into economic progress for the country, as the mills produce more goods and create more wealth. The "wheel of fortune" does indeed move faster, and it does so in both its primary and its punning sense: Mr. Tulliver's financial affairs, though slowly entangled, can suddenly plunge him into bankruptcy when the fatal lawsuit is lost. And of course Mr. Tulliver is especially vulnerable because he cannot keep up—his wheel, the mill wheel at Dorlcote Mill, does not depend on steam.

Mr. Deane's view of a progressively accelerating world has dangerous implications for a gradualist narrative. If the rate of economic change has increased so much in just forty years, then perhaps current rates are *not* adequate to explain all phenomena. The Victorians were certainly not reluctant to make this projection, and elements of *The Mill* testify to the power of such a view. Although his prospects are initially poorer than those of his brothers-in-law, Mr. Deane rises, through his position in trade, to a higher social and economic level. Tom is able to accelerate the payment of his father's debts because he speculates in Laceham goods rather than simply hoarding his accumulated wages. But Eliot's gradualist vision requires that such rosy views of rapid progress be problematized and limited. Thus the system which promotes the fortunes of a Mr. Deane does the same for Wakem and also produces a Stephen Guest. Tom pays off the family's debts in such a single-minded way that he estranges himself from the very family he is trying to save. And the machinery of the St. Ogg's wharves literally goes too fast for Maggie and Tom, overtaking and sinking their boat.

On some occasions the narrator intervenes directly to correct the nongradualist views of certain characters. She claims a wider perspective, an acquaintance with childhood and adulthood, with the beginning and the end of the story, that enables her to place such limited, local beliefs into a larger and more meaningful frame. As Tom imagines how he will imitate the success of his Uncle Deane, he "leaped over the years . . . in the haste of strong purpose and strong desire," but the narrator reminds us that he "did not see how they would be made up of slow days, hours, and minutes" (3, 5). At the opposite extreme lies Tom and Maggie's childhood belief that life will always be the same. Defending the reality and bitterness of their childhood grief, the narrator explains that "if we could recall that early bitterness, and the dim guesses, the strangely perspectiveless conception of life that gave the bitterness its intensity, we should not pooh-pooh the griefs of our children" (1, 7). It is this "strangely perspectiveless conception of life" that exaggerates the emotions of childhood: pain is more painful, happiness seems like it will last forever. Without devaluing these emotions, the narrator attempts to locate them within her historical sense of succession and cycle, of the linear movement of individual human life against the seasonal repetitions of the natural world. She tells us that although Tom and Maggie were wrong to think their lives wouldn't change, "they were not wrong in believing that the thoughts and loves of these first years would always make part of their lives" (1, 5). The memories of these first years will alter, will take on new and different meanings, but they will not simply be lost and forgotten, in part because of the connecting bond of the earth, "where the same flowers come up again every spring that we used to gather with our tiny fingers as we sat lisping to ourselves on the grass—the same hips and haws on the autumn hedgerows—the same redbreasts. . . ." It is not surprising that in the midst of this commentary, another strongly Wordsworthian moment when two historical points are being linked together, the narrator again dissolves the present of the narrative into the moment of narration. The "mild May day" of Tom and Maggie's childhood becomes the mild May day of the narrator, for whom the beauties of nature around Dorlcote Mill evoke "the subtle inextricable associations" which "the fleeting hours of our childhood left behind them" (1, 5).

## NONPROGRESSIONISM

Lyell's opposition to progressionism in the organic world is bluntly stated in the *Principles*: "the progressive development of organic life, from the simplest to the most complex forms, . . . though very generally received, has no foundation in fact" (1:145). Darwin, while accepting the general movement of life from simple to complex in the *Origin*, denied any necessary progression in the evolution of species. An index of Lyell's reluctance to abandon his position entirely can be seen in the interpretation he offers of Darwin's theory even when he came to accept it publicly: "It may be thought almost paradoxical that writers who are most in favour of transmutation (Mr. C. Darwin and Dr. J. Hooker, for example) are nevertheless among those who are most cautious, and one would say timid, in their mode of espousing the doctrine of progression. . . . [O]ne of the principal claims of Mr Darwin's theory to acceptance is, that it enables us to dispense with a law of progression as a necessary accompaniment of variation."[31]

*The Mill* does contain what appears to be an endorsement of progressionism, at least in human society. After having remarked on the "sordid life" and the "oppressive narrowness" of the "emmet-like Dodsons and Tullivers," the narrator discusses how this narrowness acts on lives like Tom's and Maggie's:

> it has acted on young natures in many generations, that in the onward tendency of human things have risen above the mental level of the generation before them, to which they have been nevertheless tied by the strongest fibres of their hearts. The suffering, whether of martyr or victim, which belongs to every historical advance of mankind, is represented in this way in every town and by hundreds of obscure hearths: and we need not shrink from this comparison of small things with great; for does not science tell us that its highest striving

is after the ascertainment of a unity which shall bind the smallest things with the greatest? In natural science . . . there is nothing petty to the mind that has a large vision of relations, and to which every single object suggests some vast sum of conditions. It is surely the same with the observation of human life. (4, 1)

Despite the reference to "the onward tendency of human things" and the successively rising mental level of each generation, the passage also speaks of the "strong fibres" which tie a generation to its predecessors. And the passage is embedded in a chapter that discusses the deep traditionalism and religious paganism of the Dodsons and Tullivers. The appeal to science, the unity of the smallest things with the greatest, does not really support claims of progress but of interrelation. Suffering is not incompatible with advance, but it is difficult to see how, for martyr and victim, suffering can be redeemed by assertions that it is an inevitable part of progress. Indeed, the novel dramatizes the fact that suffering problematizes progress, that the "onward tendency" of human knowledge does not necessarily equal progress.[32]

Given that Eliot was a proponent of the Development Hypothesis and a supporter of Darwin but resistant to the equation of "evolution" with "progress" so common in the work of Comte, Spencer, and Lewes, we should not be surprised when Sally Shuttleworth demonstrates that most of the evolutionary references in *The Mill* call progress into question, especially where notions that man is more advanced than apes, or that members of a civilized society are more advanced than savages, are concerned.[33] The difference between Aunt Pullet and a Hottentot is not so great as it may seem, and these "emmet-like Dodsons and Tullivers" (4, 1) are not a noble breed. This questioning of progress and progressionism, however, far from repudiating Lyell and Darwin, offers a particular reading of evolutionary theory with a highly Lyellian coloring—where progression is not a "necessary accompaniment" of variation. To give some idea of the complexities here, I will consider an example not discussed by Shuttleworth.

In the scene where Bob Jakin explains to Maggie how he cheats his shrewish female customers by marking the end of a yard of fabric with the far side of his thumb but then cutting on the near side, the narrator says that Bob's thumb represented "a singularly broad specimen of that difference between the man and the monkey" (4, 3). The reference is almost certainly to Richard Owen's *On the Gorilla* (1859), which used the thumb to emphasize the uncrossable gulf between human and apes and to celebrate man's position at the apex of creation. (One of the Bridgewater Treatises was devoted to the hand as an example of man's divine creation as a physically perfect being.) Owen was a friend of Eliot and Lewes, and they admired his work (Lewes dedicated *Sea-Side Studies* to him), but he was also the implacable opponent of Darwin. His gorilla lecture invokes the arguments of the natural theologian that man is a separate and specially created being who has not evolved from the apes. "[N]ine-tenths . . . of the differences . . . distinguishing the gorilla and the chimpanzee from the human species," says Owen, "must stand in contravention of the hypothesis of transmutation and progressive development"—the gap between man and monkey is simply too wide to admit the development of one into the other.[34] Elsewhere, Owen suggests that evolutionary ideas are unworthy even of refutation: "As to the successions, or coming in, of new species, I might speculate on the gradual modifiability of the individual; on the tendency of certain varieties to survive local changes, and thus progressively to diverge from an older type; on the production and fertility of monstrous offspring; . . . on the probability of . . . a variety better adapting itself to the changing climate or the conditions than the old type . . . but to what purpose?"[35] And if Owen's opposition to evolution by natural selection is obvious, so too is his disdain for the non-progressionism of uniformitarian geologists. Extinct mammals, Owen argues, have been replaced by "more numerous, varied, and higher organized forms of the class," and "so far . . . as any general conclusion can be deduced from the large sum of evidence," it is "against the doctrine of the Uniformitarian."[36]

Although Owen's quiet presence in *The Mill* seems to endorse progressionism without evolution, background and context suggest a very different conclusion. Owen's position in the gorilla lecture—the natural theologian criticizing Lyell's rejection of progressionism but strongly denying that progression has occurred through evolution—is virtually the opposite of Eliot's, who accepted evolution but wondered if evolutionary change could be called progress. Indeed, in a letter to Owen in 1867, Lewes specifically mentions their long-standing disagreement with Owen's natural theology: "that we differ profoundly respecting Design and the Creator is an old story—*that* difference never yet disturbed our harmony" (*Letters*, 8:407). But Owen's views are invoked in *The Mill* only to be rejected. The narrator's assertion of the difference between man and monkey is another example of Eliot questioning claims of progress, for Bob uses that difference to cheat his customers. Man's reason and moral sense were supposed to set him apart from animals; to Darwin's opponents, natural selection could not possibly account for the development of either. Bob, despite his physical superiority over the monkey, puts his reason to use for immoral ends. The scene is comic, and we want to agree with Bob that he only cheats those who try to cheat him—just as later we will enjoy the way he cheats Aunt Glegg in the chapter where she "Learns the Breadth of Bob's Thumb"—but it is *Maggie* who disapproves of Bob's action and calls it cheating. As the irony dissolves the difference between man and monkey, it suggests the very thing that Owen abhorred, that monkey may have evolved into man. If the dogmatic anti-evolutionary nonprogressionism of Lyell is left behind, more so is Owen's anti-evolutionary form of progressionism with its natural theological trappings. Man and monkey are brought closer together, simultaneously offering the possibility of evolutionary change but denying that such change always or inevitably constitutes "progress." Owen is used to bring his scientific enemies into the text, to defend the Lyellian side of Darwin, evolutionary development without necessary progression.

## ACTUALISM

Of all the events in the novel, it is the concluding flood that most challenges the uniformitarian basis of the narrative. The mere existence of a flood suggests catastrophism, for here is a violent force unlike any other in the novel, suddenly altering the face of the story as it sweeps over the face of the earth. Moreover, the flood evokes images of the Noachian deluge, that linchpin in the controversy over Genesis and geology. If we must account for the presence of Richard Owen in the interstices of this novel, then we must surely account for the presence of a flood which, according to Shuttleworth, provides a final vindication of catastrophe theory.

From Bulwer-Lytton and Henry James on, critics of *The Mill* have complained about the flood. Even those who have defended Eliot have felt that the novel's foreshadowings and image patterns do not adequately equip us to deal with the flood's sudden impact on the narrative. In an early essay, Barbara Hardy sees a certain appropriateness in the pacing of the novel's final books and the sudden, superficial attraction of Maggie and Stephen, but she admits that "[e]ven a well-prepared *deus ex machina* like this one may give the appearance of an arbitrary concluding rush."[37] In a more recent essay, however, she criticizes Eliot's formal devices as too artificial and inadequate: "The flood," she concludes, "is the Providence of the novel."[38]

Providence. Deus ex machina. Deluge. Diluvial wave. Any flood in English literature is bound to raise biblical associations, but critics use these terms because the novel itself sanctions them: the flood appears almost in answer to Maggie's prayer and is "that awful visitation of God," a "great calamity" (7, 5). Tom looks at Maggie and infers "a story of almost divinely-protected effort" (7, 5). And the novel's only direct reference to a specific geologist is not to Lyell or Darwin but to Buckland—catastrophist and natural theologian par excellence.

When Lucy, Maggie, and Stephen discuss the upcoming meeting of the Book Club of St. Ogg's, Stephen recommends that Lucy choose Southey's *Life of Cowper,* "unless she were inclined to be philosophical, and startle the ladies of St. Ogg's by voting for one of the Bridgewater Treatises." Lucy is curious about these "alarmingly learned books," and Stephen is willing to enlighten both her and Maggie, as it is "always pleasant to improve the minds of ladies by talking to them at ease on subjects of which they know nothing." Stephen then becomes "quite brilliant in an account of Buckland's Treatise, which he had just been reading." The moment is one of strong attraction between Maggie and Stephen: as she becomes "absorbed in his wonderful geological story," he becomes "fascinated" by her "clear, large gaze." When his recollections begin to "run rather shallow," Stephen offers to bring the book to Maggie, which Lucy playfully forbids (6, 2).

The passage has an autobiographical element: *The Mill's* precise historical chronology reveals that Maggie's introduction to Buckland comes in roughly the same year (1839) and at the same age (20) as Eliot's, and Maggie, like Eliot, is fascinated by the arguments of natural theology and the efforts to reconcile Genesis with geology. This interest sets Maggie apart from her fellow townspeople as well as from Lucy, all of whom are unaware of or uninterested in Buckland's book, but her interest is also fraught with danger. Stephen's eloquence enables him to appear to advantage, and his knowledge of Buckland's "wonderful geological story" makes him desirable to Maggie for reasons similar to those behind her attraction to Philip. Stephen possesses knowledge of a world where the apparently chaotic is divinely ordered, where even the trivial has its place and suffering has a purpose—a narrative of destruction and redemption in the natural world. Stephen revels in this position of having power over Maggie, of her interest in something only he can impart, for he sees himself "as if he had been the snuffiest of old professors, and she a downy-lipped alumnus" (6, 2). Ironically, although Lucy denies Stephen access to Maggie through Buckland's book, she replaces it, unwittingly, with other means: " 'I must forbid your plunging Maggie in books. I shall never get her away from them; and I want her to have delicious do-nothing days, filled with boating, and chatting, and riding, and driving'" (6, 2). Plunging Maggie into boating is, as it turns out, far more dangerous than plunging her into books, and as if to underscore the point, Stephen responds to Lucy's command by proposing a row on the river to Tofton that very moment.

However "wonderful" Buckland's "geological story" seemed to the George Eliot of 1841, the George Eliot of 1859 had written scathingly about the assumptions of both scriptural geology and natural theology. By associating Buckland's treatise with Stephen Guest's verbal energy and eloquence, Eliot makes it clear that Buckland's arguments, like Stephen's, must be repudiated in spite of the temptations involved. This monument of catastrophic geology, of the doctrine of progressionism by special creations, and of natural theology, is connected to the flood of *The Mill* and the flood of Noah by its language of "streams" and "plunging," by its juxtaposition with a discussion of rowing and the flux of the tides. But it is allowed into the text only for the purpose of being repudiated and controlled by the novel's relentlessly uniformitarian current.

"Uniformitarianizing" Buckland, however, can be fully achieved only in the context of uniformitarianizing the novel's flood, and this Eliot attempts to do in several ways. At the most basic level, this is the purpose of the extensive web of foreshadowings, historical references to floods, and river imagery. The flood, as we have long known, was in Eliot's plans from the beginning, but the reader must be prepared if she is to resist the catastrophist interpretation that Lyell and Darwin, in the case of real floods, knew was so easily made.

The novel's concluding chapter also works to place the flood in a local, uniformitarian context. Eliot considered the novel's principal "defect" to be the rapid pacing of the final volume in comparison with that of the first two, the "want of proportionate fullness" in the third volume versus the "epische Breite" (epic breadth) of its predecessors.[39] She admitted composing the final chapters in an unusual burst of inspiration and perhaps sensed that her flood had too much the appearance of a universal deluge, a catastrophic event, for the Conclusion emphasizes the *limitations* of the flood. Moving ahead five years, the narrator looks at the flood and its effects from a wider perspective and finds that its desolation had little lasting effect on much of the natural and human worlds: the flood "had left little visible trace on the face of the earth" and "the wharves and warehouses on the Floss were busy again" (7, Conclusion). Dorlcote Mill had been rebuilt and Dorlcote churchyard had "recovered all its grassy order and decent quiet." Life has returned to normal for most people, continuing very much as if nothing had ever happened.

But we are still in a Lyellian world where changes in nature are neither completely cyclical nor completely directional. Although the flood leaves little visible trace, there *is* some trace, and because there has been death, restoration is not complete. However unimportant the deaths of Tom and Maggie are to the rest of the community, they are not forgotten by Stephen, Lucy, and Philip. The same is true of destruction in the natural world: "Nature repairs her ravages—but not all. The uptorn trees are not rooted again; the parted hills are left scarred; if there is new growth, the trees are not the same as the old, and the hills underneath their green vesture bear marks of a past rending. To the eyes that have dwelt on the past, there is no thorough repair." The narrator's verdict is that floods, however destructive to individual lives, are part of a uniformitarian model of history that includes social as well as geological phenomena. When the floodwaters recede, the natural and human worlds again reveal the fusion of cycle (the corn grows, the wharves are busy, the mill wheel turns) and sequence (there is a new tomb in the churchyard).

Again, the language of geology intrudes. Nature, which appeared to Tom and Maggie in childhood as an endless and perfectly replicating cycle, only appears so. The flood leaves behind "scars" and "marks," but they are visible only to those, like the geologist or the novelist, of trained vision, those with eyes that have dwelt on the past and on the present. Like Lyell, the narrator denies pure cycle because pure cycle would deny history by eliminating the ability to differentiate among various historical periods and various human lives. But the reconstruction of the geological narrative reveals the flood to be a local uniformitarian event just as the reconstruction of the human narrative reveals that it, too, lies within that same local uniformitarian context.

This contextualization is perhaps better understood by considering the narrator's famous comparison between the ruined villages along the Rhone and the crumbling castles of the Rhine. Eliot connects the story of the Rhone villages with her own narrative of the Dodsons and Tullivers: the villages are "dismal remnants of commonplace houses, which in their best days were but the sign of a sordid life," and they oppress the narrator with the feeling that much of human life "is a narrow, ugly, grovelling existence," language that echoes comments about the "sordid life" of the Dodsons and Tullivers and the "oppressive narrowness" experienced by Tom and Maggie (4, 1). By way of contrast, the era of the "castled Rhine" was "a time of adventure and fierce struggle," "a day of romance" dominated by violent robber-barons who represent "demon forces" and are like "the wild beast" in their "tearing and rending." Although the narrator admits that "these Rhine castles fill me with a sense of poetry," she abjures that story for the more prosaic tale of sordid narrowness in which there are "no sublime principles, no romantic visions" (4, 1).

The deserted Rhone villages declare, however, not that they gradually sank into oblivion, but that "the swift river once rose, like an angry, destroying god, sweeping down the feeble generations whose breath is in their nostrils, and making their dwellings a desolation" (4, 1). The realistic narrative of narrow, sordid, everyday lives ends, like *The Mill on the Floss* itself, with a flood. The biblical language of this passage, with its direct quotation from the account of the Noachian deluge in Genesis 7:22, parallels the biblical language used at the end of the novel. Yet in both cases a flood is brought within the uniformitarian perspective, and *the* Flood is indirectly contextualized within a uniformitarian rubric. Lyell, as we have seen, discredited catastrophism as a form of romance, as a geological narrative lacking realism and therefore only superficially creative; in Eliot's "machinery of Fact," both the flood that destroys the Rhone villages and the flood that kills Maggie and Tom are the culminations of uniformitarian and realist narratives, not of catastrophist romances.[40]

## PROBLEMS: LOCALISM AND LITERARY TRADITIONS

We have seen Darwin's efforts in the *Journal* to reconcile the localist traditions of British geology with a global narrative of Humboldtian proportions. There is a similar tension existing in the literary traditions which infuse *The Mill on the Floss*. On the one hand, Eliot's novel is relentlessly local, a detailed depiction of the people and things in the immediate vicinity of St. Ogg's. Its pastoral and Wordsworthian elements, its attention to everyday events and everyday characters, help to make it, in common with *Adam Bede,* a Natural History of English Life similar to Riehl's social history of the German people. But like Middlemarch if less so, St. Ogg's is part of a wider social web. It is connected to the outside world through business and commerce, its mercantile economy sustained by the river which "links the small pulse of the old English town with the beatings of the world's mighty heart" (4, 1), from the bustling wharves at St. Ogg's and the Dutch vessel that carries Maggie and Stephen to Mudport, to Tom's speculations in Laceham goods and the upriver irrigation that threatens to reduce Mr. Tulliver's water power.

Maggie is herself associated with both the love of familiar things, the claims of family, duty, and the past, and the desire to discover that wider, bustling world, to free herself from the limitations that familiar things can impose. "What novelty," the narrator asks, "is worth that sweet monotony where everything is known, and *loved* because it is known?" (1, 5), yet even the very young Maggie is fascinated by books of natural history and geography, books that "tell you all about the different sorts of people in the world," about countries full of creatures other than horses and cows (1, 4). But if Maggie's monotony, even in childhood, is often not sweet, novelty can be equally bitter: trips beyond her geographical boundaries—whether running away to the gypsies, traveling to King's Lorton, or riding in the boat with Stephen—are usually associated with trouble and suffering. Even the commerce which serves some characters so well becomes the cause of Tom and Maggie's death when their boat is capsized by the fragments of machinery dislodged from the wharves.

Maggie's difficulty in negotiating her way between local demands and global desires parallels Eliot's difficulty in making of Maggie Tulliver's life a story of "epische Breite" and classical tragedy despite the narrow and restricted circumstances of that life. Lyell and Darwin eschewed the catastrophic event of global proportions—the geological equivalent of the tragic and the epic—in favor of the quiet change generated by gradual, local forces, but they sought to make these forces capable of effects both tragic and epic.[41] Eliot attempts a similar feat, but the more successful she is at dramatizing the narrowness and sordidness of Maggie's life, the more difficult it is for the reader to sustain a vision of her life as meaningful on a tragic or epic scale.[42] Indeed, Eliot's efforts to transform Maggie's uniformitarian biography are frustrated by the fact that Maggie's tragedy is "of that unwept, hidden sort, that goes on from generation to generation, and leaves no record" (3, 1)—the only evidence of this unfossilized tale lies in the memory of the narrator.

## PROBLEMS: THE ADEQUACY OF ACTUALISM

Henry Adams, recalling his efforts to write an article, under Lyell's guidance, about the tenth edition of the *Principles*, remarks that he had difficulty accepting Lyell's explanation of ice-age glaciation as a uniformitarian phenomenon. "If the glacial period were uniformity," asks Adams, "what was catastrophe?"[43] Although Eliot's uniformitarian narrative casts the flood as an example of actualism, generations of readers and critics have not been wrong to apply Adams' question to the flood in *The Mill*.

*The Mill on the Floss* is a novel in which the narrator controls perspective through her manipulation of recollection and memory. Within the narrative of the flood itself, events depend on the historical perspective of the characters, on the ability of the people of St. Ogg's to recall and interpret their own past, in which floods have always played a prominent role. The legend of Ogg, the narrator tells us, "reflects from a far-off time the visitation of the floods, which even when they left human life untouched, were widely fatal to the helpless cattle, and swept as sudden death over all smaller living things" (1, 12). This legend offers, as with Niebuhr and the legends of early Rome, a catastrophist vision of history in which violent floods are an everyday occurrence; the "true" history of St. Ogg's can be obtained only by recognizing that these floods have been compressed into a catastrophic narrative and then by reconstructing them into a sequence of similar events separated by years of gradualist change. In St. Ogg's, the floods of the past, even those that are part of local legend, confirm the validity of actualism, for despite their romantic and mythical qualities, they are no different in degree or kind from those of the present. The flood at the end of *The Mill*, although it is a unique historical event, recapitulates crucial elements of some of these earlier floods: death of livestock, human misery but minimal loss of life, destruction of Dorlcote Mill.

The same is true of more recent floods. During Tom and Maggie's childhood, for example, there was a spring when "the meadows was all one sheet o' water" (1, 6). Tom tells Bob and Maggie of an earlier "big flood" when "the sheep and cows were all drowned, and the boats went all over the fields ever such a way" (1, 6). This is the flood witnessed only by the old men, who, during the rains at the end of the novel, "talked of sixty years ago, when the same sort of weather, happening about the equinox, brought on the great floods, which swept the bridge away, and reduced the town to great misery" (7, 5).[44] Yet the townspeople, especially the young, ignore the warnings of the old men. As a people who "did not look extensively before or after" and "inherited a long past without thinking of it" (1, 12), they prefer to rely on more recent evidence that seems to confirm what they want to believe, that violent floods are a thing of the past. For the previous generation, "[t]he Catholics, bad harvests, and the mysterious fluctuations of trade, were the three evils mankind had to fear: even the floods had not been great of late years" (1, 12); the young assume the trend will continue:

"threatenings of a worse kind, from sudden thaws after falls of snow, had often passed off in the experience of the younger ones; and at the very worst, the banks would be sure to break lower down the river when the tide came in with violence, and so the waters could be carried off, without causing more than temporary inconvenience . . ." (7, 5).

Eliot's emphasis on the townspeople's wishful thinking indicates that the flood marks not an intervention from the catastrophic past, but a continuation of actualist forces. Large floods have always been part of life in St. Ogg's: those of the distant past were not greater; those of the present, properly understood, have not diminished. The residents do not need to take the perspective of geological time to understand that such floods constitute a "cause now in operation"—they need look back only as far as the human lifetimes of their living neighbors. Their refusal to do so is elaborated by Eliot early in the novel, well before the flood, in geological language: "Since the centuries when St. Ogg with his boat and the Virgin Mother at the prow had been seen on the wide water, so many memories had been left behind, and had gradually vanished like the receding hill-tops! And the present time was like the level plain where men lose their beliefs in volcanoes and earthquakes, thinking to-morrow will be as yesterday, and the giant forces that used to shake the earth are for ever laid to sleep" (1, 12). Shuttleworth concludes from this passage that the inhabitants of St. Ogg's are "firm uniformitarians" and that Eliot is here implicitly criticizing the uniformitarianism of both Lyell and Darwin.[45] But the passage begins with the uniformitarian image of memories gradually vanishing in the fashion of slowly eroding hills. And those people, living on the level plain, who "lose their beliefs in volcanoes and earthquakes," who think "to-morrow will be as yesterday," and who believe "the giant forces that used to shake the earth are for ever laid to sleep," are not uniformitarians, but catastrophists. It was the catastrophists who argued that there must have once been giant geological forces greater than anything we see today. It was the catastrophists who believed that these forces had diminished over time, and that this progressive diminution had made the earth habitable for man. Buckland's language is worth recalling here: "an uninhabitable planet, through long successions of change and of convulsive movements" has been brought to "a tranquil state of equilibrium, in which it has become the convenient and delightful habitation of man." The catastrophists' earth was the scene of violent change in the *past*, not in a convenient and delightful present; "thinking to-morrow will be as yesterday" is the stasis of Buckland's "tranquil state of equilibrium," not the dynamism of Lyell's steady-state but always changing earth. The inhabitants of St. Ogg's merit the opprobrium directed by Lyell against those who adopt this erroneous view of the earth's "quiescence," those who "refuse to conclude that great revolutions in the earth's surface are now in progress" (*Principles*, 1:313).

In his emphasis on our ignorance of existing agents of geological change, Lyell complains that catastrophists always assume that forces are progressively diminishing. Yet, says Lyell, even if we should discover "by unequivocal proofs" that certain forces were once "more potent instruments of change over the surface of the earth than they are now," it would be "more consistent with philosophical caution to presume, that after an interval of quiescence they will recover their pristine vigour, than to regard them as worn out" (1:165). The flood in *The Mill* vindicates the uniformitarian memories of the old men who correctly predict that "quiescence" is illusory and will give way to "pristine vigour," not the catastrophist hope of the rest of the people that the power of the floods is "worn out." Only in an unrealistically brief frame of reference—brief even by the standards of human history—can the flood be viewed as a violation of actualist principles.

Yet Lyell's expandable definition of uniformity can be unsatisfying when it seems to render the term meaningless, and Henry Adams' question suggests that the problem lies in the elasticity of the uniformitarian conception of actualism. To explain phenomena like the ice age, Lyell's actualist principle that the earth's history could be accounted for solely by reference to "Causes Now in Operation" had to be loosened. In Page's formulation, quoted by Lewes, "[t]he agencies that now operate on and modify the surface of the Globe . . . are the same in kind, though differing it may be in degree, as those that have operated in all time past."[46] This qualification—"though differing it may be in degree"—shows that uniformitarianism did not so much defeat catastrophism as absorb it, for "causes now in operation" had come to mean that present forces are analogous, rather than strictly equivalent, to earlier forces. Yet this had once been Buckland's position in his Bridgewater Treatise (1:528).

Unhappiness with the elasticity of actualism can perhaps be felt more intensely in a narrative like Eliot's, drawn in human rather than simply geological proportions. In Lyell's text of theoretical geology, the destructive effects of earthquakes are less important than their constructive effects in balancing erosion: though they are "so often the source of death and terror to the inhabitants of the globe" and though they "fill the earth with monuments of ruin and disorder," earthquakes are "a conservative principle in the highest degree, and, above all others, essential to the stability of the system" (*Principles,* 1:479). As we have seen, Darwin was overwhelmed by the effects of a massive Chilean earthquake, but in his published work he used the quake to confirm Lyell's uniformitarian theory of mountain building. For Eliot, however, in a novel where the principal focus is on the hidden tragedy of human lives, an event like the flood cannot be uniformitarianized without seeming to devalue the very lives the narrator has been at pains to unearth and expose to sympathetic view. The extraordinary destructive power of Darwin's Andean stream would have been harder to accept as a uniformitarian phenomenon if, like the Chilean earthquake or the flood at St. Ogg's, its victims had included human beings as well as rocks and stones.

In a notebook entry a decade or more after *The Mill,* Eliot expressed some uneasiness with actualist assumptions, at least when human history is at stake: "Is the interpretation of man's past life on earth according to the method of Sir Charles Lyell in geology, namely, on the principle that all changes were produced by agencies still at work, thoroughly adequate & scientific? Or must we allow, especially in the earlier periods, for something incalculable by us from the data of our present experience? Even within comparatively recent times & in kindred communities how many conceptions & fashions of life have existed to which our understanding & sympathy has no clue!"[47] Like the passage about the memories of the residents of St. Ogg's, Eliot's criticism of uniformitarianism contains a uniformitarian observation: since "the data of our present experience" are inadequate to account for "comparatively recent times" and "kindred communities," why should we assume that they can account for ancient times and disparate cultures? Eliot is not rejecting uniformitarian geology here, but she *is* questioning whether Lyell's actualist methodology is "thoroughly adequate" for the sociological and anthropological investigations of "man's past life on earth." There are times when the "fashions of life"—and we can't help but recall the fashions of Aunt Glegg's clothes chest—differ so much in degree, as to amount to a difference in kind.

*The Mill on the Floss,* then, bears witness to uniformitarianism's absorption of catastrophism even as it reveals, perhaps against its will, uniformitarianism's desire to have it both ways, to claim that an event like the flood is both a source of death and terror, *and* a conservative principle essential to the stability of geological and narratological systems. Although the novel champions uniformitarianism against Buckland's wonderful geological story and offers a Lyellian reading of the "progress" in progressive development, it almost inevitably leaves us questioning, with Henry Adams, the adequacy of Lyell's actualism. If readers and critics—and Eliot herself—have expressed dissatisfaction with the ending of the novel, it may have as much or more to do with the tensions within the uniformitarian model Eliot was appropriating than with the awkward intrusion of the novelist's providential designing hand.

# 5

# Ruskin's "Analysis of Natural and Pictorial Forms"

In turning to Ruskin, we come to a more conservative voice in the nineteenth century's methodological debates. Ruskin's lifelong love of science was of course intimately connected with his love of art, particularly landscape art. As his father, worried that the public might find the geology of the fourth volume of *Modern Painters* "a deviation from the right path," put it, "[f]rom Boyhood he has been an artist, but he has been a geologist from Infancy, and his geology is perhaps now the best part of his Art, for it enables him to place before us Rocks and Mountains as they are in Nature."[1] The first of Ruskin's many articles on geology and mineralogy appeared in the *Magazine of Natural History* when he was fifteen; as an undergraduate at Oxford he studied under William Buckland and made diagrams and sketches to accompany Buckland's geological lectures. Even when his interests turned more directly to the criticism of art and society, Ruskin remained abreast of contemporary science, applying it in his work, injecting himself into its controversies, and critiquing its place in the culture.

At the base of Ruskin's approach to both art and science was a methodology of induction: precise, unbiased observation leading to certain knowledge. This methodology, articulated forcefully in the initial volume of *Modern Painters* (1843), remained essentially consistent throughout his career. By itself, such an approach comes dangerously close to the naive Baconianism that was being repudiated by his contemporaries. But Ruskin, too, took care to distinguish copying or imitation, the objective observation that effaces the observer, from the Wordsworthian phenomenologist concerned with rendering the natural world as it appears to the individual observer. Moreover, in volume 2 of *Modern Painters*, he erected upon this inductive foundation a theory of the imagination, for without imagination there can be no great art. Thus the broad methodology developed in *Modern Painters* was consistent with the scientific methodology being articulated during the same period, a method combining the inductive observation of facts with the operation of the imagination, with hypotheses and hunches and guesswork. The difference was Ruskin's even stronger emphasis on induction as the foundation for the work of the imagination, and his insistence that the prevailing sin of art had almost invariably been its futile attempt to rise to the level of the imagination without regard for the real.

The evils of art for Ruskin were, in John Rosenberg's words, "the two extremes of mindless imitation and false idealization," the one epitomized by Dutch realism, the other by Old Masters such as Claude Lorrain and Gaspard Poussin.[2] The corresponding evils of science were materialism and speculation. Materialism was the more obvious evil: it looked carefully at nature, but it did so without regard either for nature's relationship to human beings or for its moral and spiritual truths. It reduced nature, and especially human interaction with nature through sight, sound, emotion, and thought, to atoms and numbers, forces and chemical reactions. To correct this error required a wider focus that included the response of the observer. Especially later in his career, Ruskin responded with increasing hostility to the evil of speculation, for John Tyndall's concept of "the scientific use of the imagination" threatened to claim the imagination in the service of materialist science. Ruskin had already made extensive use of poetry and myths as sources for the historical component of the human response to natural phenomena, but by employing them heavily in his own works of natural history on flowers, birds, and rocks he was also able to provide an example of the proper use of the scientific imagination. Just as important but largely unnoticed, however, is that his attacks on Tyndall (and other scientific materialists) for their failure to ground speculation in careful observation were an effort to expose the improper use of the scientific imagination. His objections to specific scientific theories were thus directed not only—and perhaps not even primarily—at content but at methodology, and those objections constitute a reaction against the prevailing winds of methodological change. Ruskin wanted a Turner for modern science, and Tyndall wasn't it.

In *The Queen of the Air* (1869), his essays on "The Greek Myths of Cloud and Storm," Ruskin critiques most of the major scientific theories of his day: wave theories of sound and consciousness, kinetic theories of matter and heat, Huxley's protoplasm and Darwin's natural selection. This critique emphasizes two pitfalls of modern science: "the baseness of mere materialism on the one hand, and . . . the fallacies of controversial speculation on the other" (19:351). Writing in the year prior to Tyndall's British Association address on the scientific imagination, Ruskin focuses on "mere materialism," arguing that such explanations are incomplete or inadequate precisely where our interest is greatest: in the interaction with the human. Sound waves may cause the eardrum to vibrate, but "my hearing is still to me as blessed a mystery as ever" (19:353). Thoughts may be conveyed by brain waves, but "consciousness itself is not a wave. It may be accompanied here or there by any quantity of quivers and shakes, up or down, of anything in the universe that is shakeable—what is that to me? My friend is dead, and my—according to modern views—vibratory sorrow is not one whit less, or less mysterious to me, than my old quiet one."

To restore the human element, Ruskin approaches the study of natural phenomena here through the study of myth. Myths, like materialist explanations, are rooted in careful observation of the natural world, but they express the human response to that world in emotional and moral terms. Thus, commenting on John Tyndall's recent Royal Institution lecture, "On Chemical Rays, and the Light of the Sky," Ruskin contends that Tyndall's experiments, which showed that the blue of the sky is caused by colorless, infinitesimal particles, merely confirm the "instinctive truth" articulated by the Greeks in the symbolism associated with Athena: "the bright blue of the eyes of Athena, and the deep blue of her aegis, prove to be accurate mythic impressions of natural phenomena which it is an uttermost triumph of recent science to have revealed" (19:292).

Ruskin takes a similar position three years later in his Oxford lectures entitled "The Relations of Natural Science to Art," published as *The Eagle's Nest*. The natural history of anything properly consists of three branches: the "poetry of it," "the actual facts of its existence," and "the physical causes of these facts" (22:245). By poetry, Ruskin means the traditions and myths associated with the object or organism. This knowledge, he says, is the most important for the artist, "so that we may know what the effect of its existence has hitherto been on the minds of men." It is far more crucial than knowledge of the material causes, of the chemical and physical laws of matter that make the thing what it is, and even more than of the "actual facts" obtained through observation and description of the thing in its actual state. When Ruskin contrasts two girls looking through a telescope, the one acquainted with astronomical laws and the other familiar only with "the traditions attached by the religions of dead nations to the figures they discerned in the sky," he finds a nobler wisdom in the knowledge of the second girl than in the material analysis and mensuration of the first (22:143).

But careful observation of the external appearances of things is crucial to both wise science and wise art. In his opening lecture, Ruskin declares "the three great occupations of men," defined broadly but including their more restricted senses as well, to be Science, Art, and Literature (22:125). There are wise and foolish versions of each of these three occupations, but Ruskin denies that the distinction between them depends on the method of reasoning employed. Rather, bad logic itself "proceeds from people's having some one false notion in their hearts, with which they are resolved that their reasoning *shall* comply" (22:131). That "false notion" can of course be the exclusive reliance on materialist assumptions, but more generally it is any theory adopted in advance of observation. "It is to be wished," says Ruskin, that scientists, even those of the current century, "had done less work in talking, and more in observing" (22:124).

Returning to a distinction made in *Modern Painters III*, Ruskin explains that in depictions of nature, the graphic artist's "office" is "to show her appearances" and his "duty" is "to know them. It is not his duty, though it may be sometimes for his convenience, while it is always at his peril, that he knows . . . the *causes* of appearances, or the essence of the things that produce them" (22:209). For such mimetic purposes, the less causal science an artist knows, the better. In lecturing to students, Ruskin consistently stresses this point. Here, he does so by using an anecdote about Turner:

> He was one day making a drawing of Plymouth harbour, with some ships at the distance of a mile or two, seen against the light. Having shown this drawing to a naval officer, the naval officer observed with surprise, and objected with very justifiable indignation, that the ships of the line had no port-holes. "No," said Turner, "certainly not. If you will walk up to Mount Edgecumbe, and look at the ships against the sunset, you will find you can't see the port-holes." "Well, but," said the naval officer, still indignant, "you know the port-holes are there." "Yes," said Turner, "I know that well enough; but my business is to draw what I see, and not what I know is there." (22:210)

When you can do this, Ruskin tells his students, "when you are quite fearless of your faithfulness to the appearances of things," *then* you may "allow yourself the pleasure . . . of learning what ships or stars or mountains are in reality" (22:211). And, indeed, if the artist's purposes are more than graphic, such knowledge of causes and essences then becomes necessary (22:213).

Wise science, because it is rooted in the observation of appearances and includes the human element, avoids the fallacies of "controversial speculation" and "mere materialism." Referring back to *Modern Painters I*, Ruskin contends that the subjects of wise art and wise science will be the same, for wise art, in its efforts to represent "the likeness of the thing itself," is "only the reflex or shadow of wise science" (22:151).

## THE METHOD OF *MODERN PAINTERS*

In the preface to the final volume of *Modern Painters*, Ruskin acknowledged what he considered healthy "oscillations of temper" and "progressions of discovery" (7:9) over the course of seventeen years and five volumes. Nonetheless, he also declared that "[i]n the main aim and principle of the book, there is no variation, from its first syllable to its last." This is certainly true of the book's methodology. Ruskin's phenomenology, what he called in volume 3 "the science of aspects," is rooted in realism and mimesis, in careful observation and the accurate depiction of what one sees. Ruskin distinguishes this inductive base, even in volume 1, from "mere" imitation, which can be seen as the artistic analogue to the "mere materialism" or naive Baconianism of the scientist, for it is only on this base that the imagination can build successfully. Where the error associated with the Theoretic faculty, that faculty concerned with moral perception and the appreciation of beauty, is to limit it to aesthetics, "a mere operation of sense," the error associated with Imagination, Ruskin says in volume 2, is to "exhibit things as they are *not*" (4:35–36). This error of imagination is the artistic analogue of the scientist's controversial (because not grounded in observation) speculation.

This overarching methodology is articulated quite early in *Modern Painters I*. According to Ruskin, the landscape painter "must always have two great and distinct ends": "the first, to induce in the spectator's mind the faithful conception of any natural objects whatsoever; the second, to guide the spectator's mind to those objects most worthy of its contemplation, and to inform him of the thoughts and feelings with which these were regarded by the artist himself" (3:133). For Ruskin, great landscape art begins with "the faithful conception of natural objects." The artist's emotional and intellectual response to the landscape is of no value without this careful representation of visible nature: "[A]lthough it is possible to reach what I have stated to be the first end of art, the representation of facts, without reaching the second, the representation of thoughts, yet it is altogether impossible to reach the second without having previously reached the first. . . . And this is the reason why, though I consider the second as the real and only important end of all art, I call the representation of facts the first end; because it is necessary to the other and must be attained before it. It is the foundation of all art" (3:136). Ruskin does not limit art to "the representation of facts." This would be mere imitation, which produces pleasures "the most contemptible which can be received from art" (3:101). Great art, on the contrary, demands "the representation of thoughts," the presence of an active artistic imagination selecting and re-arranging its subject matter. But such selection and manipulation must be founded on the faithful representation of the facts of nature.

Ruskin's emphasis in the first volume on the representation of facts is also related to his strategy for rebutting Turner's critics on their own ground. These critics attacked Turner's landscapes as unfaithful depictions of nature; unlike those of the Old Masters, Turner's were subjective and impressionistic renderings full of fabulous colors, impossible lighting effects, and fanciful scenery. Ruskin, in contrast, argues that the works of the Old Masters seem to be faithful to nature only because we ourselves have not observed nature closely. If we take the trouble to do so, we discover that Turner is not less but more faithful to nature, that Turner looks at nature with greater care than any landscapist living or dead. Thus, over and over again, Ruskin stresses that seeing must actively involve both the senses and the mind: "unless the minds of men are particularly directed to the impressions of sight, objects pass perpetually before the eyes without conveying any impression to the brain at all; and so pass actually unseen, not merely unnoticed, but in the full clear sense of the word unseen" (3:142). This passivity of vision on the part of painters leads to unfaithful representations of nature that, in the hands of equally passive critics, become the standard for truth. Of Italian skies, for example, Ruskin writes:

> How many people are misled, by what has been said and sung of the serenity of Italian skies, to suppose they must be more *blue* than the skies of the north, and that they see them so; whereas the sky of Italy is far more dull and grey in colour than the skies of the north, and is distinguished only by its intense repose of light. . . . And what is more strange still, when people see in a painting what they suppose to have been the source of their impressions, they will affirm it to be truthful, though they feel no such impression resulting from it. Thus . . . they will affirm a blue sky in a painting to be truthful, and reject the most faithful rendering of all the real attributes of Italy cold or dull. (3:144)

For Ruskin, a "faithful rendering" of natural objects involves the representation of what the artist actually sees, which in turn provides access to the "real attributes" of nature. So insistent is Ruskin on this point that, like Turner in the anecdote about not painting the ship's portholes, he warns painters to avoid the common error of "suppos[ing] that they see what they know" to exist, and then "of painting what exists, rather than what they can see" (3:144–45). The representation of natural objects must be founded on close observation uncluttered by the reason or imagination, by what one knows or what one hopes to be there. Certain pine trees, for example, are dark green, but when seen from a distance their color may be tinted with purple. To be faithful to nature, the landscapist must paint them dark green if he is close enough to see them that way, but he must paint them with a purple tint if he is painting from a distance. Just because he knows the trees are dark green when seen up close does not mean he should paint them that way. As Patricia Ball puts it: "[S]eeing is for [Ruskin] the prime qualification of the great artist. Emotion and further enlightenment depend upon the completeness of the encounter with the object. If it is only vaguely seen, manipulated for the artist's convenience, or reduced to his preconceived assumptions about it, there will be no revelation, no noble emotion. Turner is to be acknowledged as a hypersensitive eye before he can be hailed as nature's interpreter. That role arises out of his subjection to the snowstorm, the cloud or the sea; he does not begin from his emotions about these things. The facts come first."[3]

In keeping with both his methodology and his argument, Ruskin devotes the bulk of *Modern Painters I* to a demonstration of Turner's faithful renderings of nature, first in the more general truths of tone, color, and space, and then in the specific truths common to landscapists of skies, clouds, rocks, earth, mountains, water, and vegetation. As an observer of nature, Ruskin is extraordinarily precise and detailed, offering pages of description of natural phenomena. In his discussion of one of the "common and general optical laws which are to be taken into consideration in the painting of water" (3:508), for example, Ruskin describes the various reflections that can be seen in different parts of the rippled surface of a body of water:

> If water be rippled, the side of every ripple next to us reflects a piece of the sky, and the side of every ripple farthest from us reflects a piece of the opposite shore, or of whatever objects may be beyond the ripple. But as we soon lose sight of the farther sides of the ripples on the retiring surface, the whole rippled space will then be reflective of the sky only. Thus, where calm distant

water receives reflections of high shores, every extent of rippled surface appears as a bright line interrupting that reflection with the colour of the sky.

When a ripple or swell is seen at such an angle as to afford a view of its farther side, it carries the reflection of objects farther down than calm water would. Therefore all motion in water elongates reflections, and throws them into confused vertical lines. The real amount of this elongation is not distinctly visible, except in the case of very bright objects, . . . whose reflections are hardly ever seen as circles or points, which of course they are on perfectly calm water, but as long streams of tremulous light. (3:506)

This "optical law" is the result of detailed observation. Such observations may be inadequate for assessing Turner's greatness as "the painter of light" (3:303n)—Ruskin says that he must postpone discussion of Turner's light effects until after his discussion of the beautiful—but they are definitely adequate for demonstrating the falsehood in the paintings of the Old Masters. In his critique of Claude Lorrain's "Sea-piece, with a Villa," for example, Ruskin complains that Claude's representation of the sun's reflection is simply wrong:

The sun is setting at the side of the picture, it casts a long stream of light upon the water. This stream of light is oblique, and comes from the horizon, where it is under the sun, to a point near the centre of the picture. If this had been done as a license, it would be an instance of most absurd and unjustifiable license, as the fault is detected by the eye in a moment, and there is no occasion or excuse for it. But I imagine it to be an instance rather of the harm of imperfect science. Taking his impression instinctively from nature, Claude usually did what is right and put his reflection vertically under the sun; probably, however, he had read in some treatise on optics that every point in this reflection was in a vertical plane between the sun and the spectator; or he might have noticed, walking on the shore, that the reflection came straight from the sun to his feet, and intending to indicate the position of the spectator, drew in his next picture the reflection sloping to this supposed point. . . . (3:511)

Ruskin is willing to give Claude the benefit of the doubt: the error is probably the result not of "license" but of "imperfect science," and imperfect science is itself the result of careless observation or of theorizing without observation. Indeed, Ruskin goes on to remark that Claude's error is "plausible enough to have been lately revived and systematized" (3:511n) in books on geometric optics and perspective, and his brief refutation of this error is instructive.

Admitting that he lacks the space to enter into a lengthy disputation, Ruskin notes that

reasoning is fortunately unnecessary, the appeal to experiment being easy. Every picture is the representation, as before stated, of a vertical plate of glass, with what might be seen through it drawn on its surface. Let a vertical plate of glass be taken, and wherever it be placed, whether the sun be at its side or at its centre, the reflection will always be found in a vertical line under the sun, parallel with the side of the glass. The pane of any window looking to sea is all the apparatus necessary for this experiment; and yet it is not long since this very principle was disputed with me by a man of much taste and information, who supposed Turner to be wrong in drawing the reflection straight down at the side of his picture, as in his Lancaster Sands. . . . (3:511n)

As is typical of him, Ruskin is hostile not to science but to what he sees as foolish science. Claude has been led astray by those who have reasoned without looking, so Ruskin reverses the process, rejecting reasoning in favor of a simple visual experiment. And this exercise in careful observation reveals that Turner's work, though seemingly at odds with both nature and optics, is in fact in closer accord with both than Claude's.

The theories of beauty and imagination developed by Ruskin in *Modern Painters II* build on this "downright statement of facts" (3:138) emphasized in the first volume. His famous recollection in *Praeterita* of drawing the Norwood ivy captures this relationship nicely. After making a careful drawing of this bit of ivy, Ruskin says he saw that he had "virtually lost all my time since I was twelve years old, because no one had ever told me to draw what was really there!" (35:311). Yet it was drawing what was really there that unleashed for him the *beauty* of the natural world; prior to that day in 1842 his drawing had provided him with "records of places," but he had "never seen the beauty of anything, not even of a stone—how much less of a leaf!" Since the Theoretic faculty's perception of beauty in external nature is modified by the Imagination in bringing that perception to the canvas, attention to natural fact is the essential starting point for the artist.

Although he complains that the commonly entertained definition of the imagination is indistinct and thus cannot be contrasted to his own with certainty, Ruskin does know that he sees little evidence of imagination in much of the art praised for being imaginative. Once again, there are two extremes to be avoided: "unpalliated falsehood and exaggeration" and "minute and mechanical statement of contemptible details" (4:286–87). Downright statement of facts—noble, not contemptible facts—is insufficient for great art, but the artist who employs imagination does so by embracing rather than spurning the facts of nature. Even the painter of the meanest imaginative power may do grand things, says Ruskin, if he will keep to strict portraiture of nature, and, indeed, "it would be well if all artists were to endeavour to do so, for if they have imagination, it will force its way in spite of them, and show itself in their every stroke; and if not, they will not get it by leaving nature, but only sink into nothingness" (4:243).

Ruskin distinguishes three forms of imagination: Associative, Penetrative, and Contemplative. He devotes most of his attention to the first two, which are present to some degree in most artists but to a high degree in very few. In all three of its forms the imagination operates mysteriously and inexplicably. Even the Associative Imagination, which Ruskin admits can sometimes be hard to differentiate from the more mechanical power of composition, instantly sees that the component ideas of a painting, though separately wrong, will be right together, functioning as an organic whole greater than the sum of its parts. Like Wordsworth and Shelley before him and T. S. Eliot after him, Ruskin expresses this instantaneous synthetic vision of the imagination by comparing it to a chemical reaction (4:234–35). Whereas the unimaginative artist foolishly adheres to laws of artistic composition—his tree trunks lean first one way and then another because that is how tree trunks "ought" to lean—the imaginative artist knows but is not restrained by the laws of nature, and these laws are not a restraint because "[t]hey are his own nature" (4:239). The imagination is not to be measured by its distance from nature. Anything that looks unnatural or appears to be false can contain no imagination (4:247).

It is through the Penetrative Imagination, however, that the mind reaches the greatest truths. "Intuition and intensity of gaze" result in "a more essential truth than is seen at the surface of things" (4:284). As with the truth of the Associative Imagination, this "Imaginative Verity" is distinguished by Ruskin from "falsehood on the one hand, and from realism on the other": "The power of every picture depends on the penetration of the imagination into the TRUE nature of the thing represented, and on the utter scorn of the imagination for all shackles and fetters of mere external fact that stand in the way of its suggestiveness" (4:278). In Tintoretto's version of the Massacre of the Innocents, for example, the terror of the scene is not depicted directly in an act of butchery, but suggested by the painting's visionary light. In his *Entombment of Christ*, the manger is brought to Jerusalem "that it may take the heart to Bethlehem" (4:278). Nonetheless, such imaginative liberties remain grounded in the truths of nature even as they reach after spiritual and moral truths: the imagination "must be fed constantly by external nature" (4:288). In the service of these higher truths, "we may build up the mountain as high as we please," says Ruskin, "but we must do it in nature's way, and not in impossible peaks and precipices" (4:312).

This inductive foundation of the imagination enables us to clarify the relationship in *Modern Painters III* between "the science of aspects" and "the science of essence," and to appreciate Ruskin's return in *Modern Painters IV* and *V* to the beauties of mountains, leaves, and clouds. George Landow says that the "long disquisition" on geology in volume 4 induces the "bored and bewildered" reader to "glance at the title page to reassure himself that he is still pursuing a work about Turner."[4] John Rosenberg calls Ruskin's scientific studies "irrelevant," contending that the discussion of truth in *Modern Painters* is simply an "excuse" for tacking these studies onto a more important aesthetic theory.[5] John James Ruskin knew that readers might respond this way to his son's geology, yet he proclaimed the centrality of that geology for establishing the faults of most landscapists, and, implicitly, the successes of Turner. The science of *Modern Painters* simultaneously offers a corrective for both bad science and bad art.

In the penultimate chapter of volume 3, "The Moral of Landscape," Ruskin argues that "scientific pursuits" have "a tendency to chill and subdue the feelings, and to resolve all things into atoms and numbers" (5:386). While science is to be praised for raising us from "inactive reverie" to "useful thought," it impedes most minds from rising to "higher contemplation." Like Keats complaining about the "cold philosophy" that "unweaved the rainbow," Ruskin questions "whether any one who knows optics, however religious he may be, can feel in equal degree the pleasure or reverence which an unlettered peasant may feel at the sight of a rainbow" (5:387). Also like Keats, however, Ruskin directs his complaint not against science but against *materialist* science, the science that resolves all things into atoms and numbers.

It is at this point that Ruskin distinguishes the science of aspects from the science of essence. The science of aspects is phenomenological: it studies the appearances of nature as observed by a human beholder. The science of essence, on the other hand, studies the physical causes of natural phenomena. In Ruskin's oft-quoted words, "there is a science of the aspects of things, as well as of their nature; and it is as much a fact to be noted in their constitution, that they produce such and such an effect upon the eye or heart (as, for instance, that minor scales of sound cause melancholy), as that they are made up of certain atoms or vibrations of matter" (5:387). Yet what is often not noticed is that Ruskin introduces the phrase "science of aspects" not to condemn the science of essence for its lack of beauty and emotion, but to defend his phenomenological approach to nature, with its interest in the aesthetic (or, as Ruskin would say, theoretic) and emotional response of the observer, as scientific. Although he clearly feels that modern science has become so dominated by materialist assumptions about vibrating atoms that science and materialism can virtually be equated, Ruskin does not contend that the science of essence is necessarily materialist. Rather, materialism is the perversion of the science of essence. Ruskin's announced aim is to determine "the exact relation between landscape-painting and natural science, properly so called" (5:384), and this involves the recognition that there are "good and evil forms of this sympathy with nature" (5:385). To find moral or spiritual truth in material nature, the science of aspects, is the good form of this sympathy; to see nothing in a leaf but vegetable tissue is not natural science, properly so called. Both the ditch-water realism of the Dutch school and the materialism of modern science can be avoided or corrected by the science of aspects.

By the same token, the science of aspects is the antidote to the idealized falsehood of the Old Masters and the controversial speculations of science. This is evident in the paragraph immediately following, where Turner is declared the master of the science of aspects, Bacon the master of the science of essence. The point of the comparison, of course, is not to disparage Bacon but to elevate Turner. At the end of the previous chapter, Ruskin had argued that "Turner, the first great landscape-painter, must take a place in the history of nations corresponding in art accurately to that of Bacon in philosophy;—Bacon having first opened the study of the laws of material nature, when, formerly, men had thought only of the laws of the human mind; and Turner having first opened the study of the aspect of material nature, when, before, men had thought only of the aspect of the human form" (5:353). Bacon overthrew the speculations of Scholasticism for the study of the essence of material nature, but modern science has perverted that study into mere materialism. Turner reversed the Renaissance's tendency to set beauty above truth, a propensity that had in fact *banished* beauty from depictions of both natural and human subjects, "from the face of the earth, and the form of man" (5:324), and turned instead to the aspects of material nature, where real beauty is to be found.

Bacon also stands at the head of the list of human thinkers whose "dreaming love of natural beauty" is subordinate to the love of "result, effect, and progress" (5:359–60). Ruskin acknowledges that the figures in this list, which also includes Newton, Milton, and Richardson, are of higher dignity than corresponding figures (e.g., Byron, Shelley, Keats) whose love of natural beauty is intense. Since he is defending the virtues of the contemplation of landscapes, however, Ruskin contends that the subordination of the love of nature is never to the individual's advantage, and that while the love of nature does not guarantee moral or intellectual excellence, it "is an invariable sign of goodness of heart and justness of moral *perception*" (5:376). The lover of nature is the person who, standing before a group of trees with four other men, including an unfeeling artist and an engineer obsessed with examining the fibers of the trees' roots, sees "the thing itself" (5:358). His view of the parts will not be as detailed, but he will see the whole most clearly; his vision will encompass all that the others see. Ruskin uses a passage of tree description from Wordsworth to exemplify his point, and yet he immediately characterizes "the chief narrowness of Wordsworth's mind" as his inability to understand "that to break a rock with a hammer in search of crystal may sometimes be an act not disgraceful to human nature, and that to dissect a flower may sometimes be as proper as to dream over it" (5:359). Ruskin had himself broken many a rock and dissected many a flower, but he had done so without losing his sense of the beauty, and he wanted modern scientists, the inheritors of the Baconian revolution, to regain the ability to work in the same spirit.

We can see these same patterns and tensions in the discussion of mountain beauty in volume 4, and particularly in the discussion of "Turnerian topography." For Ruskin, "topography" is the art of relating the facts of landscape simply and straightforwardly; "Turnerian topography" is the art of relating those same facts imaginatively (6:27). Topographical painting is neither easy nor mechanical, and it requires the selection of pleasing and instructive subjects from which ugly, interfering details of human construction (in Ruskin's examples, prominent new tourist hotels that spoil the panoramas of Swiss towns) have been removed. But otherwise the topographical painter is to treat his painting as a mirror, reflecting the natural scene with utter accuracy and without alteration. The Turnerian topographer, on the other hand, paints not "the actual facts" of a landscape but "the impression it made on his mind" (6:32). This impression comes from the action of the Penetrative Imagination; it is a sudden, instinctive, and organic vision, not to be confused with manufacture by abstract "laws" of composition that demand lights, colors, and forms be placed in certain fixed relations. For Ruskin, the simplest of all artistic canons, and one of the few that is inviolable, is that "it is always wrong to draw what you don't see" (6:27). The topographical painter, who sees only things that exist, must paint these "actual facts." The Turnerian painter, blessed with imagination, sees things that do not exist, or that do not exist apparently, but must still paint this "impression on his mind." If, says Ruskin, "when we go to a place, we see nothing else than is there, we are to paint nothing else. . . . If, going to the place, we see something quite different from what is there, then we are to paint that—nay, we *must* paint that, whether we will or not; it being, for us, the only reality we can get at. But let us beware of pretending to see this unreality if we do not" (6:28). Pretending to see unreality is the false imagination of Claude and his modern disciples, and Ruskin again stresses that for him the truly imaginative landscapist is also a skilled topographer: "before [painters] dare so much as to *dream* of arranging [nature], they must be able to paint her as she is: nor will the most skilful arrangement ever atone for the slightest wilful failure in truth of representation: and I am continually declaiming against arrangement, not because arrangement is wrong, but because our present painters have for the most part nothing to arrange. They cannot so much as paint a weed or a post accurately; and yet they pretend to improve the forests and mountains" (6:39–40n).

To demonstrate this relationship, Ruskin compares a topographical drawing, executed by himself, of the Pass of Faido, with a drawing of the same site made by Turner. The scene itself, Ruskin admits, is not particularly interesting or impressive: the mountains are neither high nor striking in form, the heaps of stones in the valley are arranged in a disagreeable clutter, and the gently winding road appears smooth and safe. For the traveler, however, the sensation produced by this scene is much different. Having descended into this valley from even higher peaks through one of the narrowest ravines in the Alps, the traveler sees the scene in terms of the one he has just passed through: the gorge is narrow and terrible, the stones are emblems of the river's fury, and the road coils precariously by overhangs and precipices strewn with the remnants of avalanches. Simple topography, the physical facts, cannot capture this impression in the spectator's mind. Turnerian topography can, but it does so not by simply abandoning the physical facts.

Turner's primary alteration is to expand the scale of the scene. To render the beholder's state of mind, he makes the landscape more like the landscape the traveler has just experienced. In his drawing, the mountains are higher and steeper, and the river's violence is captured by removing some trees and a bridge and by making its banks higher and more rocky. Ruskin sees Turner's alterations in this drawing as faithful to the impression made on Turner's mind, a perfect example of the "mental chemistry" (6:41) of the Associative Imagination, instantaneously and instinctively making the adjustments and combinations necessary to depict the truth perceived by the Penetrative Imagination. Yet these alterations are also faithful to how the scene would have appeared had it actually existed. Turner had built up mountains as high as he pleased, but he did it in nature's way, not in impossible peaks and precipices.

For Ruskin, the imagination is not "a false and deceptive faculty," but "the most accurate and truth-telling faculty which the human mind possesses" (6:44). Turnerian topography thus renders higher truths than simple topography—the "real" and "most important facts" of "the world's outside aspect" are not to be found in "maps, nor charts, nor any manner of mensuration" (6:45). And this is true in a wider sense: in comparison to imaginative truth, "all mathematical, and arithmetical, and generally scientific truth" is "truth of the husk and surface, hard and shallow" (6:44). Here again, the scientific is associated with measure and number. Its truths are limited because untouched by the imagination. But if it is possible for a science of aspects to generate a Turnerian topography, it is possible for it to correct the perversions and limitations of a science of essence as well. It was this that Ruskin tried to do in his own scientific works, and it was this that led him to attack John Tyndall's alternative vision of the scientific imagination.

## RUSKIN AND THE METHODS OF SCIENCE

In a paper for the Mineralogical Society nearly forty years after the appearance of *Modern Painters I*, Ruskin declared that "precisely the same faculties of eye and mind are concerned in the analysis of natural and of pictorial forms" (26:386). Ruskin's modern critics, even his most sympathetic ones, while acknowledging this similarity in approach, have tended nonetheless to describe Ruskin's science, both in content and in method, as unscientific. John Rosenberg, for example, refers to Ruskin's geological essays of the 1870s and early 1880s as "capricious" and "deliberately unscientific."[6] Rosenberg argues that Ruskin turns to such studies of nature, undertaken during a period of depression and breakdown, to ward off madness. Ruskin's geology "was a combination of lifelong study of the Alps and pleasant putterings in his kitchen with toast-crumb morraines and glaciers of blancmange. It is a measure of his increasing intellectual isolation that his master in geology was Saussure, whose *Voyages dans les Alpes* had first appeared in 1779. With defensive arrogance, he attacked the contribution of contemporaries, never quite realizing that his own was not science but play."

Frederick Kirchoff and Robert Hewison accord more status to Ruskin's "play," suggesting that Ruskin's "deliberately unscientific" methodology is in fact an alternative methodology that reacts against the prevailing methods of contemporary science. In his discussion of *Proserpina,* Ruskin's book on flowers, Kirchoff declares that Ruskin "resists defining science as 'scientific method'" because he strongly opposed a scientific way of knowing "that had depended on the increasingly restricted function of the observer and the elimination of all but quantifiable perceptions."[7] Hewison sees Ruskin as resisting the movement of mid-nineteenth-century science to new methods of "analysis and experiment."[8]

Kirchoff and Hewison are right: Ruskin's science was shaped by his methodology, and this methodology was influenced by his perception of the methods of his scientific contemporaries. But by now we can recognize that the characterization of nineteenth-century "scientific method" as objectivity and mathematization is incomplete. For Ruskin, there was always a distinction between scientific and aesthetic perception, but not an absolute difference. The science of aspects and the science of essence could never be the same, but a "natural science, properly so called"—a science of essence with a science of aspects included—was possible. Ruskin critiqued the materialism of contemporary geology, botany, and zoology in *Modern Painters; The Eagle's Nest, Deucalion, Proserpina,* and *Love's Meinie* extended this critique, but more important they responded to the challenge not of the increasingly restricted function of the observer, but of the increasingly expanding function of the observer's scientific imagination.

Ruskin's methodological comments are most extensive in his writings on geology, the science he knew most intimately. In his discussion of alpine geology in *Modern Painters IV,* Ruskin notes that "the natural tendency of accurate science is to make the possessor of it look for, and eminently see, the things connected with his special pieces of knowledge; and as all accurate science must be sternly limited, his sight of nature gets limited accordingly" (6:475). In contrast, his own approach to natural objects is to remain "as much as possible in the state of an uninformed spectator of the outside of things, receiving simply what impressions the external phenomena first induce" (6:476). "I closed all geological books," writes Ruskin, "and set myself . . . to see the Alps in a simple, thoughtless, and untheorizing manner; but to *see* them, if it might be, thoroughly" (6:476). Saussure is the only geologist whose help he has received because "all other geological writers whose works I had examined were engaged in the maintenance of some theory or other, and always gathering materials to support it. But I found Saussure had gone to the Alps, as I desired to go myself, only to look at them, and describe them as they were, loving them heartily . . . more than himself, or than science, or than any theories of science . . ." (6:476).

Saussure's verbal representations of the Alps, like Turner's pictorial representations of them, strive for careful and theoretically neutral observation of the visible. As he had said in the opening page of *Voyages dans les Alpes*, geology "must be cultivated only with the aid of observation, and systems must never be but the results or consequences of facts."[9] For Ruskin, reason and imagination are essential, but they follow this process of Baconian observation; in most nineteenth-century geology, on the contrary, they have preceded it. Thus in *Modern Painters* Ruskin utilizes his own drawings and observations of Mont Blanc to challenge the "chief theory with geologists" for its formation (6:253), while one of Turner's alpine paintings serves as the basis for a description of the laws of rock cleavage (6:269–75). In *Deucalion*, his collection of geological essays, Ruskin's stance is similar. "The theories respecting the elevation of the Alps," he writes, remain "uncertain and unsatisfactory," and therefore our own work "must waste no time on them; we must begin where all theory ceases; and where observation becomes possible,—that is to say, with the forms which the Alps have actually retained while men have dwelt among them, and on which we can trace the progress, or the power, of existing conditions of minor change. . . . I do not care . . . how crest or aiguille was lifted, or where its materials came from, or how much bigger it was once" (26:112–13). Ruskin clings to Saussure for methodological reasons, but he does so as much for Saussure's opposition to unsupported theory and hyperactive imagination as for his human love of the mountains.

Tyndall's addresses on the scientific imagination to the British Association for the Advancement of Science in 1868 and 1870 were, as we saw in chapter 1, celebrations of a methodology rooted in hypothesis, invention, and guesswork that Tyndall sought to differentiate from the half-truths of materialist explanations with which he was commonly associated. Tyndall's arguments were controversial: the *Saturday Review* responded by claiming that what we admire in the scientific process is not "the leap of the imagination" but "the splendid series of inductions verified step after step by rigorous experiment and observation," the "firmly balanced and duly graduated tread of a mind . . . careful to plant every step on the assured ground of fact or experience."[10] Despite such criticisms, "the scientific imagination" almost immediately became a cultural catch-phrase, and as such it posed a direct challenge to Ruskin's own imaginative and antimaterialist but fact-based methodology. It is no accident, I think, that Ruskin's three books of natural history began appearing in the early 1870s, a few years after Tyndall's addresses.

Like Ruskin, Tyndall was a great lover of the Alps, traveling through them frequently to study the same natural phenomena that fascinated Ruskin. He reported the results of his research, particularly of his work on glaciers, in lectures at the Royal Institution, and he wrote several books on the subject, including *The Glaciers of the Alps* (1860) and *The Forms of Water in Clouds and Rivers, Ice and Glaciers* (1872), the latter based on a series of Royal Institution lectures for children. In these books and lectures, however, Tyndall engaged in an acrimonious dispute about glacier motion with J. D. Forbes, whose own theory was first published in *Travels through the Alps* in 1843. Ruskin had met Forbes in 1844 and considered his book the definitive work on alpine glaciers, for Forbes, like Saussure, had taken the trouble to observe, and had built his theory on the foundation only of what he had seen. Indeed, Ruskin quotes Forbes copiously in his discussion of the central Alps in *Modern Painters IV,* even according him pride of place over Saussure. In the ensuing controversy, which extended well beyond Forbes's death in 1868, Ruskin was active in defense of Forbes.[11] He lectured on glaciers himself at the Royal Institution in 1863, championing Forbes over Tyndall, and repeatedly introduced the subject into his various works, including *Deucalion,* the new editions of *Modern Painters,* and *Fors Clavigera.*

At stake in this bitter and protracted dispute with Tyndall was the methodological base of Ruskin's life work in natural and pictorial forms. Tyndall was a powerful antagonist on two counts: he was a scientist who embraced the imagination while claiming to use his eyes, who employed speculation and materialism in mutually supporting ways. To counter Tyndall's claims, Ruskin used a strategy similar to the one he had adopted in defending Turner. Before he could praise Turner as a great imaginative painter, Ruskin had to show that, contrary to the critics' claims, Turner's work was true to nature. In attacking Tyndall, Ruskin could not simply remain in safety on his own antimaterialist ground; he had to show that Tyndall's scientific imagination was false because Tyndall was not a good observer.

Ruskin had long been using myth as an antidote to materialism, but in the works of natural history it also serves as an antidote to diseased imagination. In *Deucalion*—and like *Proserpina* and *Love's Meinie,* Ruskin's title immediately suggests the importance of the mythological, moral, and spiritual in the study of nature—Ruskin stresses as "a first principle" that "the feeblest myth is better than the strongest theory: the one recording a natural impression on the imaginations of great men, and of unpretending multitudes; the other, an unnatural exertion of the wits of little men, and half-wits of impertinent multitudes" (26:99). In this passage, Ruskin uses Darwinism as his example of "theory," but later he extends the criticism to "any other conceivable materialist theory" (26:336). The problem is a failure of imagination: the "half-wits" have no "imaginative power" and foolishly pride themselves on this, while little men like Tyndall exert their wits unnaturally. Inseparable from this improper use of the imagination is of course the failure to begin with careful observation, and Ruskin does not exempt himself from this failure. Recalling his declaration in *Modern Painters I* that in painting snowcovered peaks the artist must abide by the dictum that "there must be on every high summit as much snow as it can carry" (3:447), Ruskin notes: "There *must*, I said. That is the mathematical method of science as opposed to the artistic. Thinking of a thing, and demonstrating,—instead of looking at it. Very fine, and very sure, if you happen to have before you all the elements of thought; but very dangerously inferior to the unpretending method of sight. . . . If I had only looked at the snow carefully, I should have seen that it wasn't anywhere as thick as it could stand or lie" (26:179–80).

The unpretending method of sight is the proper foundation for hypothesis and deduction, especially for students. Ruskin thus found Tyndall's misleading and imprecise analogies, so central to the scientific imagination, particularly dangerous. He emphasizes that *Deucalion, Proserpina,* and *Love's Meinie* are intended to provide "an absolutely trustworthy foundation" in natural history for students of science and art by "sifting what is really known from what is supposed" (26:197). Ruskin's students are absolutely forbidden from "draw[ing] any logical inferences from what they know of stalagmite, to what they don't know of chalcedony. They are not to indulge either their reason or their imagination in the feeblest flight beyond the verge of actual experience" (26:205). The "securely accumulated" evidence resulting from such restraint will later facilitate the inductive establishment of "general conclusions . . . not as 'theories,' but as demonstrable laws" (26:206).

Ruskin consistently bemoaned premature efforts to explain unobserved or unobservable phenomena while observable phenomena remained shrouded in obscurity. This was one of the reasons behind his frequent criticisms of microscopes and telescopes. Indeed, Ruskin contends in *Deucalion* that science's impatience with observing—and drawing—the observable is creating a widening separation of the artist from the man of science (26:116). Comparing sketches by both Forbes and Tyndall of the Mer de Glace glacier, Ruskin sarcastically offers the most damning criticism he can make: that "Professor Tyndall is unable to draw anything as seen from anywhere, . . . such inability serving farther to establish the sense of his proud position as a man of science, above us poor artists, who labour under the disadvantage of being able with some accuracy to see, and with some fidelity to represent, what we wish to talk about" (26:161).

Tyndall's inaccurate vision and unfaithful representations, Ruskin later argues, are both a product of, and a contributor to, bad theories and bad explanations. In a discussion of the formation of ice on the surface of calm water, Ruskin notes that the hexagonal structure of the ice crystals is "sufficiently explained, in Professor Tyndall's imagination, by the poetical conception of 'six poles' for every hexagon of ice" (26:358). In *The Forms of Water,* Tyndall had claimed this as a triumphant example of the scientific use of the imagination: "Our first notions and conceptions of poles are obtained from the sight of our eyes in looking at the effects of magnetism, and we then transfer these notions and conceptions to particles which no eye has ever seen. The power by which we thus picture to ourselves effects beyond the range of the senses is what philosophers call the Imagination, and in the effort of the mind to seize upon the unseen architecture of crystals, we have an example of the 'scientific use' of this faculty."[12] Tyndall contends that by seeing the effects of magnetism he is then able to explain the form and cause of the unseen ice crystals; by analogy, each hexagonal crystal thus consists of six "poles." Ruskin, of course, sneers at such an explanation: "Perhaps!—if one knew first what a pole was, itself—and how many, attractive, or repulsive, to the east and to the west, as well as to the north and the south—one might institute in imaginative science—at one's pleasure;—thus also allowing a rose five poles for its five petals, and a wallflower four for its four, and a lily three, and a hawkweed thirteen" (26:359). Tyndall is no Turner. He can't draw because, unlike Turner, he leaps too rapidly from observation to imagination, and his vague, theoretical, metaphorical language gives him away.

In *Ruskin and the Art of the Beholder,* Elizabeth Helsinger argues that Ruskin's theories of perception sought to correct, on the one hand, the scientist's failure to take into account his response to nature, and on the other hand, the poet's tendency to see nature subjectively, exclusively through his own emotional state—what Ruskin christened the "pathetic fallacy." According to Helsinger, the Ruskinian "beholder" is a "wandering natural scientist" who, "like [the] visionary artist, does not merely observe but perceives: his own responses, shaping his observations, are an admitted part of his subject."[13] Applying Helsinger's argument to Tyndall, himself a "wandering natural scientist," we can see that the threat he poses to Ruskin is not simply in his materialism, but in a materialism supported by the scientific imagination. Tyndall's scientific use of the imagination results in the scientific version of the pathetic fallacy.

In *Modern Painters III*, Ruskin describes the poet who succumbs to the pathetic fallacy as one "who perceives wrongly, because he feels, and to whom the primrose is anything else than a primrose: a star, or a sun, or a fairy's shield, or a forsaken maiden" (5:209). The language, including the floral example, recalls the passage above in which Ruskin sneers at applying the term "poles" to a crystal of ice or the petals of a flower. Tyndall's failure, in Ruskin's eyes, is not that he lacks an emotional, imaginative response to nature but that this response is too strong. His language expresses admiration for the "architecture" of the ice crystal and reverence for the science that can explain such natural beauties by going "beyond the range of the senses." This emotional response leads Tyndall to employ his imagination to construct a scientific explanation for the crystal. But just as the primrose is anything else than a primrose for the poet who feels, the ice crystal is anything else than an ice crystal for Tyndall—it is a magnet with six poles. In Ruskin's view, *both* poet and scientist are susceptible to the pathetic fallacy, and in both cases the result is an inaccurate perception of nature. Tyndall is dangerous because he is guilty of both "the baseness of mere materialism" and "the fallacies of controversial speculation." He is the Dutch school and Claude Lorrain, contemptible realism and idealized falsehood, rolled into one.

The methodological threat posed by Tyndall can also be measured by the shift in Ruskin's response to the wave theory of light and to kinetic theories of matter and heat. We have already seen that in *The Queen of the Air* Ruskin used Tyndall's lecture on light and the sky to show that modern science had only just proven truths that the ancient Greeks had known intuitively and expressed in their myths about Athena. Ruskin read this lecture while correcting the proofs for *The Queen of the Air* and inserted the reference to it in the preface. He called Tyndall's conclusions "an uttermost triumph of recent science" and his experiments "magic of the finest sort" and a "true wonder" (19:292). He even apologized to Tyndall and "all masters in physical science, for any words of mine, either in the following pages or elsewhere, that may ever seem to fail in the respect due to their great powers of thought, or in the admiration due to the far scope of their discovery" (19:292–93). It is tempting to see these words as ironic, especially given Ruskin's immediate caveat that he has "bitter reason to ask them to teach us more than yet they have taught" (19:293), but there is a strong element of genuineness that is confirmed by a letter to his friend Charles Eliot Norton the same day.[14] Tyndall's lecture, Ruskin tells Norton, is "marvellous, and must modify my preface a little—I have to apologise first for rude observations—and then say how wonderful this putting the sky in a bottle is; and then say—for last word—that I'll thank *them*—the men of science—and so will a wiser future world—if they'll return to old magic—and let the sky out of the bottle again, and cork the devil, *in*."[15]

Even as he acknowledges the exception, Ruskin refers to the rule: that materialists like Tyndall have not been teaching the right lessons, that science would be better served by observing the actual sky rather than manufacturing imitation skies in experimental tubes, and that human society would be better served by scientists using their tubes to cork the devil of pollution in. In his survey of physical theories in the body of the work, Ruskin is especially concerned with the conclusions of thermodynamics, extensively popularized by Tyndall during the 1860s, on the relationship between heat and motion, especially the explanation of heat, and thus of life, as the vibration of atoms and molecules. Ruskin acknowledges the "tendency of recent discovery" to see various forces (chemical, mechanical, thermal, vital, etc.) as interconvertible "modes of one force" or "finally . . . mere motion," but, "assum[ing] that science has done its utmost; and that every chemical or animal force is demonstrably resolvable into heat or motion, . . . we have yet to ask, What is heat? or what, motion? What is this 'primo mobile,' this transitional power, in which all things live, and move, and have their being? It is by definition something different from matter, and we may call it as we choose—'first cause,' or 'first light,' or 'first heat'; but we can show no scientific proof of its not being personal, and coinciding with the ordinary conception of a supporting spirit in all things" (19:356).

Tyndall's lecture contains much that is in keeping both with these comments and with Ruskin's other criticisms of science. He prefaces the discussion of his experiments with summaries of the wave theory of light and the kinetic theory of matter, stressing that "we are speaking of things which lie entirely beyond the range of the senses" but which are nonetheless "as truly *mechanical* as they would be if we were dealing with ordinary masses of matter, and with waves of sensible magnitude."[16] The lecture is filled with the kind of analogies between visible and invisible phenomena that Ruskin will later ridicule, and Tyndall celebrates the ability to duplicate natural effects in his experimental tube. When he announces that he can produce a blue sky in his tube that can "rival . . . the deepest and purest Italian sky," we cannot help but recall Ruskin's barbs in *Modern Painters I* at artists and critics who passively accept the hearsay that Italian skies are bluer than northern ones, and accept this as a standard of truth.[17]

In spite of all this, Tyndall's lecture also has much to attract Ruskin. It tries to answer the kind of simple question about observable phenomena—why is the sky blue?—what Ruskin wanted scientists to tackle. Although Tyndall supplies materialistic explanations that rely on what Ruskin would have considered "theories" about unobservable atomic and molecular behavior, his experiments provide simple, occular demonstrations of the formation of blue skies and clouds. His explanations of Italian skies and alpine sunsets, while going beyond "the range of the senses," are the result of "intellectual vision" rather than the scientific imagination.[18] Ruskin is even willing to accept the existence of the imponderable ether—the hypothetical fluid required by the wave theory of light as the medium through which light waves are propagated—on which Tyndall's explanations depend, even though this theoretical entity is established not through direct observation but through "the closest reasoning of modern physicists" (19:292). Still, what should surprise us here are not Ruskin's characteristic strictures about the limitations of Tyndall's science, but his willingness to make use of that science.

Fifteen years later, Ruskin recalled Tyndall's experiments in a similar fashion, invoking the visible demonstrations while avoiding reference to materialistic assumptions and imaginative leaps. In the early 1880s, Ruskin again became interested in cloud phenomena, reprinting the cloud sections of *Modern Painters IV* and *V* as a separate volume and delivering the two lectures at the Royal Institution entitled "The Storm-Cloud of the Nineteenth Century." The physicist Oliver Lodge, an admirer of Ruskin who was interested in the same phenomena and concerned by Ruskin's imprecise definition of clouds in the "Storm-Cloud" lectures, sent Ruskin in January 1885 a copy of his recent lecture to the British Association, "Dust," just published in *Nature*. Lodge explains that it is this "fine impalpable dust" which causes the blue of the sky and the formation of clouds and rain.[19] He refers many times to Tyndall's earlier experiments and demonstrates some of his predecessor's results using the same experimental apparatus. The lecture, originally delivered in Montreal, opens with a Ruskinian warning to Canadians about the dangers of human dust in the form of industrial pollution—"you will not realise the blessing of fresh and pure air until you have lost it"—and quotes at one point from Ruskin's "Storm-Cloud" lecture.[20]

Lodge's letter to Ruskin opened a fairly intensive two-month correspondence on these sky phenomena. Initially, Ruskin interpreted Lodge's lecture as providing an explanation of blue sky different from Tyndall's: in his first letter to Lodge, Ruskin noted that "the attribution of the blue colour of the sky to water instead of air is not only left without proof, but without reference to some marvellous results of Tyndall's a while since, in which he made small firmaments in tubes" (37:514).[21] Lodge corrected Ruskin's initial misreading of the lecture—due in part to his narrow understanding of Tyndall's earlier results and in part to Lodge's lack of clarity in explaining the relationship between his work and Tyndall's—by providing Ruskin with a further explanation of atmospheric dust. His letter included a "condensed account of the chief feature of the kinetic theory of gases—the rapid movements of the individual molecules even in stationary air," and a discussion of "the nature of evaporation and condensation, as due to the same sort of imperceptible but rapid molecular movement and interchange of particles across the superficial boundary separating air and water" (37:517n). In his response, Ruskin proclaimed himself "still in great molecular agitation . . . at the *entirely* new things you have told me about perpetual motion and universal motes, and have got to accustom myself to this notion of the perpetual fidgets of calm water—and the motes even in Athena's blue eyes—the very cause of their blue!" (37:525).

Ruskin's playful language cannot obscure how unsettling "perpetual motion and universal motes" were for him. He told Lodge that he was "staggered and appalled" by the kinetic theory, "sick and giddy" at the thought of molecular motion (37:520–21). He reluctantly agreed to grant Lodge the existence of dust motes as nuclei for atmospheric condensation, but warned him that "I don't a bit believe in them yet!—except in Tyndall's experiments at the Royal Institution" (37:527). In both cases, Ruskin's difficulty lies in being asked to accept the existence of something he not only cannot see, but that is at variance with what he does see. What "visible . . . changes to mortal eyes" will these "mystic motions" produce, he asks Lodge (37:521). He "believes" in Tyndall's experiments because Tyndall produced blue sky and visible mist from dust-free air, whereas Lodge's explanations depend on infinitesimal dust particles the presence of which is assumed rather than demonstrated. "[L]et us waste no time in hypotheses," Ruskin urged

Lodge in an allusion to Newton's "hypotheses non fingo," "I never made but one in all my life, and that was wrong. I only want to know what *is*" (37:526). Ruskin is reluctant to relinquish his understanding of Tyndall's results not only because they confirm the intuitive truths of Greek mythology, but also because Tyndall in this case really did focus on "visible changes" rather than "mystic motions." Although willing to acknowledge what he had learned from Lodge in his corrections to the final edition of *Modern Painters IV* (1888), Ruskin told Lodge that he looked to him "to give me the *facts* of what I COULD *see* if I chose" (37:521). Lodge's authority prevents a direct challenge to the fidgets of calm water and the motes in Athena's eyes, but Ruskin raises subtle questions about the validity of the imaginative assumptions on which such explanations rest.

In the first "Storm-Cloud" lecture, Ruskin had opened his discussion of clouds by assessing the current scientific explanations of their forms. The business of scientific people, he says, is to tell us that certain things are so, that they do happen, and then to tell us what to do if we want these things to happen or not. But, "if . . . they ever try to *explain* anything to you, you may be confident of one of two things,—either that they know nothing (to speak of) about it, or that they have only seen one side of it—and not only haven't seen, but usually have no mind to see, the other" (34:17). Ruskin uses Tyndall as an example of this failure, claiming that when "Professor Tyndall explains the twisted beds of the Jungfrau to you by intimating that the Matterhorn is growing flat; or the clouds on the lee side of the Matterhorn by the wind's rubbing against the windward side of it,—you may be pretty sure the scientific people don't know much (to speak of) yet, either about rock-beds, or cloud-beds." In the second lecture, Ruskin elaborates on the latter example by noting that his own description of these "lee-side clouds" in *Modern Painters V,* while "less imaginative" than Tyndall's "banner-clouds" in *Glaciers of the Alps*, is "more comprehensive" and "closer to the facts" than Tyndall's. When the subject is not Tyndall's small firmaments in tubes, Ruskin savages his opponent's spurious imagination, and in these lectures he even distances himself more directly from the physical theories on which Tyndall's experiments were based.

Toward the end of a lengthy discussion of the causes of the various colors of clouds, Ruskin complains that he is constantly "stopped dead" by the scientists' "confusion of ideas" in "using the words undulation and vibration as synonyms":

> 'When,' says Professor Tyndall, 'you are told that the atoms of the sun vibrate at different rates, and produce waves of different sizes,—your experience of water-waves will enable you to form a tolerably clear notion of what is meant.'
> 'Tolerably clear'!—your toleration must be considerable, then. Do you suppose a water-wave is like a harp-string? (34:25–26)

This comment was written at about the same time as, and takes aim at the same target (Tyndall's *Forms of Water*) as, the criticism of Tyndall's scientific imagination in *Deucalion*. Whereas in *The Queen of the Air* Ruskin was willing to make use of the ether and to overlook Tyndall's tenuous analogizing from observable to unobservable, here he is suspicious and critical, as he was with Lodge's explanation of the kinetic theory of gases. Ruskin's specific criticism is off the mark—Tyndall is not using vibration and undulation as synonyms but claiming that the vibrations of the sun's atoms produce waves in the surrounding ether—but his ultimate point is valid. Water waves are physical objects, but "the undulating theory of light is proposed to you concerning a medium [the ether] which you can neither see nor touch" (34:26). Consistent with his own methodology, Ruskin points out that his criticism does not imply a rejection of the ether because there is no positive evidence to show that the ether does *not* exist. Rather, he is an agnostic about the ether, the wave theory of light, and kinetic theories of matter: "I neither accept, nor assail, the conclusions respecting the oscillatory states of light, heat, and sound, which have resulted from the postulate of an elastic, though impalpable and imponderable ether, possessing the elasticity of air" (34:26–27). Without a firm foundation in observable phenomena, the scientific imagination can yield no certain knowledge. Tyndall treats an analogy between water waves and light waves as a description and an explanation of what light is, and hence as sufficient grounds for asserting the existence of an ether that cannot be seen, felt, or weighed.

Going to the heart of these physical theories, Ruskin also raises the question of how the sun's atoms, as "dead matter," are to start vibrating in the first place: "*You* may fall a-shivering on your own account, if you like, but you can't get a billiard-ball to fall a-shivering on *its* own account" (34:26). Ruskin was wrong here as well, but he retracted the statement in his second lecture, quoting a passage from a paper by G. G. Stokes on "epipolized" (fluorescent) light: "Nothing seems more natural than to suppose that the incident vibrations of the luminiferous ether produce vibratory movements among the ultimate molecules of sensitive substances, and that the molecules in return, *swinging on their own account*, produce vibrations in the luminous ether, and thus cause the sensation of light" (34:59).

Stokes, the Lucasian Professor of Mathematics at Cambridge, commanded great respect from Ruskin. The holder of Newton's chair, Stokes was associated neither with Tyndall's materialism nor with his scientific imagination.[22] Moreover, the paper from which Ruskin quotes, "On the Change of Refrangibility of Light," consists primarily of the sort of careful and detailed experimentation and observation that Ruskin loved. Although Stokes was thus a safer authority for Ruskin, he was also an early proponent of the wave theory, and all his work on light was based on it. Ruskin is careful, in citing Stokes on the self-induced quality of the molecular vibrations, to begin his quotation *after* Stokes's endorsement of the wave theory and to ignore entirely Stokes's claim that the wave theory is to light what the theory of gravitation is to the heavenly bodies. While willing to grant that vibration can produce undulation (and hence that scientists are not using the terms synonymously), Ruskin does not want to include himself among those whom Stokes terms "believers in the undulatory theory of light."[23] Such "believers" may not be materialists, but their claims about the ether are scientific examples of the pathetic fallacy, based not on observation of aspects but on imaginative analogies. In 1869, such claims could be appropriated in the service of the relationship between modern science and ancient mythology; by 1884, following Tyndall's championing of materialist theories as products of the scientific use of the imagination, they must be refuted or avoided.

Like his contemporaries and his Romantic predecessors, Ruskin defended an inductive methodology that contained a place for both observation and imagination. Whether in the analysis of natural or pictorial forms, Ruskin stressed the impossibility of arriving at truth without a careful description of the aspects of things while at the same time claiming that the highest truths could be reached only through the imagination. And yet Ruskin's response to Tyndall's scientific use of the imagination was close to that of the *Saturday Review*, which argued that the proper method of science is not Tyndall's "leap of the imagination," but "the splendid series of inductions verified step after step by rigorous experiment and observation," with every step planted "on the assured ground of fact or experience."

If Ruskin's methodology appears both wider and more restrictive than that of Tyndall or even of Stokes and Lodge, it is in large measure because he wanted to avoid the twin evils of materialism and speculation. The primacy he gave to the human responses to nature in poetry and myth added a large component to what counted as a Baconian "fact." What began essentially as a response to materialism, however, later became an example of the proper use of the imagination in science. Yet even this was not sufficient: scientists like Tyndall, purveying a materialism linked to the imagination, had to be attacked on their own ground, exposed as bad observers whose speculations were disconnected from "a downright statement of facts." Responses like Ruskin's and the *Saturday Review*'s suggest the methodological trap in which science found itself by midcentury: its success in publicizing its inductive triumphs had made possible such responses to its efforts to prevent that view from becoming a naive and restrictive caricature that denied the importance of imagination and intuition to the working scientist.

Ruskin knew Mill, Herschel, and Whewell and had clearly read Whewell's *History of the Inductive Sciences*. It seems likely that he had also read Whewell's *Philosophy*, Herschel's *Preliminary Discourse*, and Mill's *Logic*. Yet what is perhaps most interesting is that in his criticisms of scientific methodology, and especially in his criticisms of Tyndall's methodology, Ruskin does not invoke these writers. In the cases of Whewell and Mill it is easy to speculate why: Whewell's Kantianism would have been uncongenial to a man who repeatedly attacked German philosophy, including at the opening of the chapter in *Modern Painters III* on the pathetic fallacy; Mill, with his utilitarian political economy and Radical politics, was considered by Ruskin to be a thinker utterly deficient in imaginative power, able to see only one or two of an issue's many sides. The silence on Herschel, whose views were probably closest to Ruskin's, is more difficult to account for, although perhaps Mill's prominent claim in the *Logic* that his work owed much to Herschel's *Preliminary Discourse* was sufficient reason. Nonetheless, Ruskin went his own way, and that way was, characteristically, both in step with, and critical of, his culture's innovations. It is easy to imagine someone who opened his autobiography by declaring himself "a violent Tory of the old school;—Walter Scott's school, that is to say, and Homer's" (35:13) declaring himself a violent methodological Tory of Bacon's school. For Bacon was one of Ruskin's intellectual heroes, the initiator of a revolution in the study of "material nature" to which the Turnerian revolution in landscape art formed the complement. In attacking Tyndall, Ruskin expressed his commitment to both revolutions.

# 6

## "Euclid Honourably Shelved"

### Edwin Abbott's *Flatland* and the Methods of Non-Euclidean Geometry

During the same years that Darwinian biology was rocking British culture and kinetic theories of matter were upsetting Ruskin's stomach, a third and in some respects more unsettling revolution was launched in that supposedly most settled of sciences: geometry. In a country where a staple of education from the lower forms to the universities was the study of Euclid's *Elements*, the development of different geometries and the contention that space may not be Euclidean and three-dimensional could not help but capture public attention. Thus when Edwin Abbott, under the pseudonym "A Square," drew on this controversy as the basis for his satirical novel *Flatland* in 1884, *Science* claimed in its initial notice of the book that Abbott's work represented "a new idea, on the part of the novelist, to make the conceptions of transcendental geometry the basis for an amusing story. . . . Readers of 'Alice behind the looking-glass' will not fail to notice the resemblance of the present work to that singular play of fancy."[1] Likening *Flatland* to the *Alice* books of Lewis Carroll—another Victorian concerned with the teaching of geometry—remains common today, and yet Abbott's book is perhaps even less a mere "amusing story" than Carroll's. Indeed, the first part of *Science*'s notice hints that *Flatland* could not have been written without contemporary developments in "transcendental geometry"—specifically, in the construction and popularization of non-Euclidean and *n*-dimensional geometries. In its full review of the book, *Science* later made this connection even more explicit, arguing that *Flatland* itself served the popularizing purpose of introducing its readers to the mysteries of these higher geometries.[2]

In fact, whatever else it does, *Flatland* is very much in the midst of a matrix of voices concerned not simply with popularizing abstruse research in mathematics, but with debating the nature of space and our knowledge of it, the status of Euclid's *Elements* as a textbook, and the methodology of geometry in relation to other sciences. Another review of *Flatland*, this one in *Nature*, locates *Flatland* quite clearly within this matrix of mathematical voices:

> A few years ago a distinguished mathematician published some speculations on the existence of a book-worm "cabin'd, cribb'd, confin'd" within the narrow limits of an ordinary sheet of paper, and another writer bewailed "the dreary infinities of homoloidal space." A third remarks, "there is no logical impossibility in conceiving the existence of intelligent beings, living on and moving along the surface of any solid body, who are able to perceive nothing but what exists on this surface and insensible to all beyond it." How delighted Prof. Helmholtz will be to find, if the Flatland writer is worthy of credence, his conjecture thus verified.[3]

The "distinguished mathematician" is J. J. Sylvester, whose address to the Mathematical and Physical Section of the British Association in 1869 had appeared in *Nature* as "A Plea for the Mathematician." In this address, Sylvester defended mathematics from T. H. Huxley's charge that it is a purely deductive science and therefore essentially uninteresting. On the contrary, claimed Sylvester, "mathematical analysis is constantly invoking the aid of new principles, new ideas, and new methods . . . springing directly from the inherent powers and activity of the human mind. . . . [I]t is unceasingly calling forth the faculties of observation and comparison, . . . one of its principle weapons is induction, . . . it has frequent recourse to experimental trial and verification, and . . . it affords boundless scope for the exercise of the highest efforts of imagination and invention." As an example of his claims, Sylvester cites the recent work of Riemann in developing non-Euclidean geometry, which, he says, shows "that the basis of our conception of space is purely empirical, and our knowledge of its laws the result of observation; that other kinds of space might be conceived to exist, subject to laws different from those which govern the actual space in which we are immersed. . . . [F]or as we can conceive beings (like infinitely attenuated book-worms in an infinitely thin piece of paper) which possess only the notion of space of two dimensions, so we may imagine beings capable of realising space of four or a greater number of dimensions."[4] For Sylvester, the ability to imagine what the experience of space would be like in dimensions other than three is sufficient to establish the empirical basis of geometry—three-dimensional Euclidean geometry is not the science of space in general, but the science of the space of our experience.

In a footnote to this passage, Sylvester points to W. K. Clifford as the source for his speculations about the bookworm. Clifford, Sylvester explains, has argued that just as the two-dimensional worm on the flat page would not be aware of a third dimension even if his page were rumpled and curved, so would we be unaware as three-dimensional beings if our space—which we suppose to be homoloidal, i.e. flat and level—were rumpled or curved in a fourth dimension. Clifford, in fact, is the "other writer" referred to in the *Nature* review of *Flatland*, the person complaining about the "dreary infinities of homoloidal space." He was the first British mathematician to understand the significance of Riemann's work, and he labored both to extend it on a theoretical level and to explain it on a popular one. For Clifford, the work of Lobachevsky and his successors constituted a major revolution in the history of human thought: "what Copernicus was to Ptolemy, that was Lobatchewsky to Euclid. . . . Each of them brought about a revolution in scientific ideas so great that it can only be compared with that wrought by the other. And the reason of the transcendent importance of these two changes is that they are changes in the conception of the Cosmos."[5] In this same lecture, originally delivered at the Royal Institution in 1873, Clifford argues that such changes in our conception of the cosmos, while upsetting long and deeply held beliefs, make "the whole of geometry . . . far more complete and interesting. . . . In fact, I do not mind confessing that I personally have often found relief from the dreary infinities of homoloidal space in the consoling hope that, after all, this other may be the true state of things."[6]

*Nature*'s "third writer" is, as the passage indicates, Hermann von Helmholtz, long active in British scientific circles and a consummate popularizer. The quotation from Helmholtz is taken from his groundbreaking and controversial essay, "The Axioms of Geometry" (1870), which marked the first effort in Britain both to popularize non-Euclidean geometry and to use its results for a war on the Kantian philosophy of space.[7] "The question respecting the origin and foundation of the axioms of geometry," wrote Helmholtz, "is a question, which, as I think, may be made generally interesting to all who have studied even the elements of mathematics, and which, at the same time, is immediately connected with the highest points regarding the nature of the human understanding."[8] Ultimately at stake were the necessary truths of religion and Kantian philosophy in relation to the empirical truths of scientific knowledge. Abbott's book, as it dramatizes this debate, endorses a methodology that, instead of rejecting, as Ruskin's had, Tyndall's scientific imagination and a Coleridgean/Whewellian idealism, appropriated both. Nonetheless, Abbott, too, wanted to protect his methodology, and with it his religion, from the extremes of an empiricism that linked fact and feeling.

## NON-EUCLIDEAN GEOMETRY: THE DEVELOPMENT OF A REVOLUTION

Helmholtz's sense that the debate over the axioms of geometry would be of interest to those who had studied only the basic elements of mathematics was certainly valid. In Britain, every schoolboy learned geometry (or suffered through it, if Tom Tulliver in *The Mill on the Floss* is any indication) from Euclid's *Elements*, yet it was centuries-old dissatisfaction with Euclid's parallel postulate that finally led to the development of alternative geometries.[9]

Since John Playfair's edition of Euclid in 1795, the parallel postulate has usually been taught in a form known as "Playfair's Axiom": given a line *l* and a point *p* in the same plane, with *p* not on *l*, one and only one line can be drawn through *p* that is parallel to *l*. From Euclid's day on, however, geometers had been troubled by this postulate. It was not self-evident because it made assumptions about the behavior of lines at infinite distances, yet it could not be proven from Euclid's other postulates. In 1767, d'Alembert christened it "the scandal of geometry." Finally, in the early nineteenth century, three men, Lobachevsky in Russia, Bolyai in Hungary, and Gauss in Germany, began to experiment independently with the possibility that the parallel postulate could not be proven, that it was in fact an arbitrary assumption, and therefore that other geometries could be developed by making a different assumption about parallel lines. Both Lobachevsky and Bolyai, for example, assumed that there are an infinite number of lines that can be drawn through *p* parallel to *l*, and discovered that such an assumption, while it led to a system of geometry very different from Euclid's, did not lead to any self-contradictions or inconsistencies.

The work of Lobachevsky and Bolyai went unnoticed until the 1850s, when the posthumously published papers of Gauss, whose fame was great, revealed that he, too, had been thinking along these lines. Equally crucial, just before Gauss's death, was the 1854 lecture delivered by his young colleague Bernard Riemann, who in essence took up the assumption that through *p* there are *no* lines that can be drawn parallel to *l*. To do this, Riemann redefined the concept of a straight line. "Straightness" is an intrinsic property, a property that depends on the surface in question. On the surface of the earth, for example, a "straight" line is a geodesic—it follows the curvature of the earth's surface. Viewed extrinsically, from outside the surface, this line is not straight, yet for someone on the surface, its curvature is not directly observable—it is the shortest distance between two points. Furthermore, if we allow the geodesics to be considered straight lines, then they are not necessarily infinite. On the surface of a sphere, a straight line eventually returns to its beginning. By performing his analysis on surfaces of *n* dimensions, Riemann thus argued that Euclidean geometry is a special case, one among many possible geometries, and that while our space seems to be Euclidean, we do not know for sure.[10]

In their popularizations of non-Euclidean geometry, Helmholtz and Clifford aggressively championed geometry as an empirical science rather than the stronghold of the idealists' necessary truths. As Bertrand Russell put it in 1897: "Geometry, throughout the 17th and 18th centuries, remained, in the war versus empiricism, an impregnable fortress of the idealists. . . . The English Empiricists . . . had, therefore, a somewhat difficult task; either they had to ignore the problem, or if, like Hume and Mill, they ventured on an assault, they were driven into the . . . assertion that . . . only the perpetual presence of spatial impressions . . . made our experience of the truth of the axioms so wide as to *seem* absolutely certain."[11] Russell's martial imagery is not hyperbolic, for non-Euclidean geometry made it possible for empiricists to attack idealism, and in particular Kantian idealism, at what was thought to be its strongest point. To the question "are [the axioms of geometry] inherited from the divine source of our reason as the idealist philosophers think," Helmholtz responded decisively that "it cannot be allowed that the axioms of geometry depend on the native form of our perceptive faculty. . . ."[12] Non-Euclidean geometry quickly became a crucial feature in discussions of scientific method and epistemology.

## VARIETIES OF FLATLANDS

In discussing *Flatland*'s position relative to these popularizations of non-Euclidean geometry, it is important to establish the incredible range and versatility—up to our own day—of Flatland narratives, what one writer early in the century called "part of the metageometer's stock-in-trade for the popular exposition of his subject."[13]

Abbott's Flatland is a two-dimensional Euclidean plane inhabited by intelligent geometrical figures. These figures have no conception and no physical experience of a third dimension and hence assume that the universe is two-dimensional. The story's main character, A Square, has a dream in which he visits Lineland, a one-dimensional world inhabited by line segments. A Square is unable to convince the King of Lineland that space is actually two-dimensional: the King can't see A Square unless A Square passes through the King's linear field of vision, but since the King has no perception of lateral movement, he sees A Square suddenly appear as a point, remain stationary, and then disappear. A Square's efforts to describe the second dimension are equally futile because the King cannot conceive a direction perpendicular to the only one he experiences. The following night A Square is himself visited by a Sphere from Spaceland, but the Sphere experiences the same difficulties convincing A Square of the reality of a third dimension as A Square had convincing the King of Lineland of the reality of a second dimension. A Square cannot conceive of a direction perpendicular to Flatland, and when the Sphere passes through Flatland, A Square sees him first as a point, then as a growing circle, then as a diminishing circle, and then as a point again before he disappears entirely. Only when the Sphere lifts A Square out of Flatland and into the third dimension, from which he can see into Flatland, does A Square finally accept the existence of the third dimension. So complete is his conversion that he asks the Sphere to take him to the land of Four Dimensions, but the Sphere angrily commands him to be silent, telling him that such a land does not exist and is indeed inconceivable.

Helmholtz's strategy begins by invoking a similar two-dimensional world:

> We live in and know a space of three dimensions. But there is no logical impossibility, in conceiving the existence of intelligent beings, living on and moving along the surface of any solid body, who are able to perceive nothing but what exists on this surface, and insensible to all beyond it. Neither would it be a contradiction to suppose that such beings could find out the shortest lines existing in their space, and form geometrical notions of it, as far as it is accessible to and perceptible by them. Their space, of course, would have only two dimensions. If the surface, on which they lived, were an infinite plane, they would acknowledge the truth of the axioms of Euclid.[14]

But for Helmholtz, this version of Flatland is merely a preliminary to the more interesting case where these intelligent beings live on the surface of a sphere; such beings would develop geometrical axioms very different from ours. These new axioms would in fact be related to Riemann's version of non-Euclidean geometry: parallel lines would intersect, the sum of the angles of a triangle would be greater than 180°, similar triangles would have to be congruent. Repeating this example a few years later in *Mind*, Helmholtz argued that his "sphere-dwellers" were sufficient to prove that "geometrical axioms must vary according to the kind of space inhabited."[15]

Yet even this surface of a sphere example was only the beginning for Helmholtz. He goes on to discuss the geometry developed by intelligent beings living on the surfaces of an egg and a convex mirror, contending that if we could communicate with such beings—much as A Square communicates with the King of Lineland and the Sphere—"neither . . . would be able to convince the other that he had the true, the other the distorted relations."[16] However, says Helmholtz, we can imagine what the objects of spherical and pseudo-spherical worlds would look like to someone whose experiences of space were Euclidean but who was somehow able to enter these different worlds. Since "we can imagine such spaces of other sorts, it cannot be maintained that the axioms of geometry are necessary consequences of an *a priori* transcendental form of intuition, as Kant thought."[17] As in the case of Sylvester, Helmholtz contends that the ability to conceive or imagine analogically what other spatial experiences would be like is sufficient to establish both that space is not necessarily Euclidean and three-dimensional, and that our minds are not constituted in such a way as to limit our experience of space to Euclidean three-dimensionality.

Helmholtz's articles seem to have been the first printed source of Flatland narratives in Britain, but these narratives quickly became and long remained a staple of popularizations of non-Euclideanism. Throughout the 1880s and 1890s, the pages of *Nature* are filled with versions of them by English, Continental, and American writers. After a hiatus from about 1905 to 1920—a period when *Flatland* was out of print—Flatland narratives experienced a revival with the efforts to popularize relativity theory, which showed that space-time is indeed curved, non-Euclidean, and four-dimensional.[18]

## CONCEIVABILITY

If twentieth-century writers have happily appropriated Flatland narratives, many late-nineteenth-century writers wanted to refute the philosophical claims that Helmholtz based on his powerful non-Euclidean narratives of spheres and eggs and mirrors, and the crucial issue was conceivability. Idealists argued that geometrical truths are established by the fact that it is impossible even to conceive of an alternative—a triangle whose angle sum does not equal 180° is inconceivable. Helmholtz contended that his examples of non-Euclidean spaces showed that the conception of such an alternative was clearly possible. Most mathematicians and philosophers, however, were not convinced: do Helmholtz's examples truly enable us to conceive non-Euclidean and *n*-dimensional spaces? is the mere ability to construct non-Euclidean and *n*-dimensional geometries really tantamount to a refutation of the Kantian view of space? Even someone as sympathetic to Helmholtz as G. H. Lewes refused to accept all his conclusions. Assessing the two sides in the Helmholtz-Jevons dispute, Lewes supports the contention that non-Euclidean geometry proves that geometrical axioms are experiential, but denies that non-Euclidean geometry challenges the truth of Euclidean axioms, the Euclidean nature of real space, or the three-dimensionality of real space. For Lewes, "Imaginary Geometry" is a logical system that has no bearing on the geometry we develop from the study of real space other than to show that we have developed Euclidean geometry because our experience of space is Euclidean. "The truths of plane Geometry are not affected by the truths of spherical Geometry," writes Lewes, "nor would the Geometry of three dimensions be a whit less true if we constructed a Geometry of *n* dimensions."[19] Imaginary Geometry is true in the sense that it is logically consistent, "[b]ut what is commonly understood as Truth is something more than this. . . . [A]lthough I admit that the non-Euclidean Geometry may be thoroughly consistent and ideally true,—i.e. within the sphere of its assumptions,—I can neither admit the legitimacy of extending any of its conclusions beyond that sphere, nor the suggestion of Gauss and Helmholtz that because we can conceive a space in which its axioms would not be truths, the Euclidean Geometry is not rigorous, not necessarily true." In Lewes' view, Helmholtz has succeeded in "conceiving" a non-Euclidean space because he has used two-dimensional illustrations, and we are familiar with two dimensions. But this does not affect the necessary truth of Euclidean axioms.

Lewes makes a similar argument for *n* dimensions. Although "attempts have been made of late to demonstrate a fourth dimension in Space, the wiser heads [among whom Lewes includes Sylvester and Clifford] refuse to accept the fourth dimension as a reality, content to use it as an artifice of calculation." Lewes' inclusion of Sylvester and Clifford among the "wiser heads" is probably an indirect criticism of spiritualists who leaped from the possibility of a fourth dimension to its reality, but it wrongly diminishes the degree to which both Sylvester and Clifford insisted on the possibility that our own space may be *n*-dimensional and non-Euclidean. Indeed, for Lewes, "conceiving" such spaces is a purely logical, analytical enterprise that should not be confused with "imagining" them. Of the argument that because we can imagine a space of two dimensions we can also imagine a space of four dimensions, Lewes denies even the first part of the proposition:

I deny that we can *imagine* (though we can *conceive*) a space of two dimensions. . . . To say that possibly there may be sentient beings for whom a third dimension does not exist, is very different from saying that we can imagine, i.e. form an image, of their space. By no effort can we *divest* ourselves of our intuitions, and form a mental picture of what the universe is to different intuitions. . . . It is [also] obviously impossible to imagine this fourth [dimension], which, never having been present to Sense, cannot be revived in Imagination. The comparatively easy resource of dropping one part of our sensible experience . . . is altogether different from the task of *adding* an entirely new sensible basis. A fourth dimension, then, must always remain an artifice. . . . We cannot imagine it, we cannot believe in it as a reality.

Yet Sylvester, we recall, had claimed that it is possible not only to conceive the experiences of two-dimensional beings, but to imagine how beings of four dimensions would "realise their space." And Helmholtz clearly believed he could "form an image" of the space of his sphere- and egg- and mirror-dwellers.

Lewes' response to Helmholtz is indicative of the historical fact that the radical implications of non-Euclidean geometry were successfully contained. An important figure in this regard, though coming from a very different philosophical position than Lewes', is the influential mathematician Arthur Cayley, whose 1883 presidential address to the British Association I want to consider at some length for the light it will later shed on Abbott's position in *Flatland*. Cayley's relationship to non-Euclidean geometry was much like Einstein's to quantum mechanics: although he prepared the way for non-Euclidean geometry in Britain, he refused to accept the implications that others drew from it, and he tried to direct research away from it (unlike Einstein and quantum mechanics, successfully) and toward projective geometry, in which the interpretation of non-Euclideanism did not raise the same philosophical difficulties.[20]

The subject matter of Cayley's address would have been of great interest to Abbott: "Mathematics connect themselves on the one side with common life and the physical sciences; on the other side with philosophy, in regards to our notions of space and time, and in the questions which have arisen as to the universality and necessity of the truths of mathematics, and the foundation of our knowledge of them."[21] Locating himself in an idealist tradition that stretches from Plato to Kant to Whewell, Cayley attacks Mill and Riemann and the Flatland narratives of Helmholtz and others, arguing that mind, not experience, is the "proper foundation" of the "truths of mathematics" (11:431–32). He begins by focusing on Mill's acknowledgment in the *Logic* that geometrical axioms are not fully empirical because geometrical figures are mental constructions. Mill concedes that it is impossible to conceive a line without breadth, but this is merely the idealization of our experience with real lines. Cayley, however, challenges Mill's emphasis on physical lines by arguing that "the purely imaginary objects are the only realities . . . in regard to which the corresponding physical objects are as the shadows in the cave; and it is only by means of them that we are able to deny the existence of a corresponding physical object; if there is no conception of straightness, then it is meaningless to deny the existence of a perfect straight line" (11:433). Cayley, in effect, reverses Lewes' terms, but the results are the same: we can imagine anything, but our conceptions are limited. Euclidean geometry is purely imaginary, but it is the only way we can conceive of space, and hence its imaginary objects, not their corresponding physical ones, are the true realities. On the other hand, we can construct a non-Euclidean geometry, and we can even imagine a non-Euclidean space, but we cannot conceive it.

In his discussion of the Flatland narratives themselves, Cayley argues that we can construct non-Euclidean geometries without modifying our notion of space as Euclidean and without diminishing the truths of Euclidean geometry. Like Helmholtz, Clifford, and Sylvester before him, Cayley invites his audience to "imagine rational beings living in a one-dimensional space (a line) or in a two-dimensional space (a surface), . . . to whom, therefore, a two-dimensional space, or (as the case may be) a three-dimensional space would be as inconceivable as a four-dimensional space is to us" (11:436–37). But dwellers on the surface of a sphere the size of our earth, he says, would initially develop a Euclidean geometry because parallel lines would appear to remain parallel. With more experience of their world, however, these sphere-dwellers would discover that these experientially derived axioms were false—two parallel lines, if followed far enough, would eventually intersect. They would then develop a spherical geometry that accurately represented the space of their experience. But this would not mean that their Euclidean system was not true, merely that it applied to an ideal space rather than the space of their experience. And that the space of their experience turned out to be non-Euclidean would not imply that all space is non-Euclidean. Where Riemann and Helmholtz argued that we have a general notion of space but learn by experience that physical space is exactly or to a high degree of approximation Euclidean,

Cayley denies that we have the freedom to conceive of our physical space in a variety of ways. Physical space is "the representation lying at the foundation of all external experience." It is the shadow on Plato's cave. Flatland narratives like Helmholtz's can represent non-Euclidean space only in Euclidean terms (as the surface of a sphere) because that is the only way we can conceive of any space. The development of non-Euclidean geometries cannot change the fact that the purely imaginary conception of Euclidean space is the true reality. That we can develop non-Euclidean geometries without changing our "notion of space" suggests that non-Euclidean geometries and non-Euclidean spaces can be kept separate (11:435).

Cayley draws similar conclusions from his considerations of the conceivability of the fourth dimension. We cannot, Cayley insists, conceive of higher dimensions; "space as we conceive of it, and the physical space of our experience, are alike three-dimensional." We can, however, imagine one-dimensional or two-dimensional beings for whom two-dimensional or three-dimensional space, respectively, would be as inconceivable as four-dimensional space is to us. If the two-dimensional surface were curved, Cayley again contends that the beings on it would discover that their system of geometry ceased to be applicable to their physical space, but that they would develop a new system of geometry rather than conclude that their space is curved in a third dimension. And this would be true even if the surface were not curved: "take the case where the two-dimensional space is a plane, and imagine the beings of such a space familiar with our own Euclidean plane geometry; if, a third dimension being still inconceivable by them, they were by their geometry or otherwise led to the notion of it, there would be nothing to prevent them from forming a science such as our own science of three-dimensional geometry" (11:436–37). Notice that Cayley says nothing would prevent these beings from forming a *geometry*; developing this geometry would not, however, in Cayley's view, amount to developing a different conception of *space*. This position that non-Euclidean geometries are logical systems valid as mathematical tools but having nothing to do with the nature of real space or our experience of it was perhaps the dominant, containing interpretation of the new geometry, embraced alike by the empiricist Lewes and the idealist Cayley.

The success of this containing strategy is exemplified by the fact that Russell's discussion of geometry focuses on Cayley's projective approach, denigrating Helmholtz's "somewhat careless attempts to popularize mathematical results" and regretting that these "romances about Flatland and Sphereland—at best only fairy-tale analogies of doubtful value—have been attacked as if they formed an essential feature of Metageometry."[22] Yet Russell, like others before him, is himself drawn into a discussion of this "careless" thinking, the power of which depends significantly on the rhetorical success of the fanciful Flatland and Sphereland examples. Similarly, despite his disparagement of Abbott's book as a "joke," Russell nonetheless echoes its subtitle ("A Romance of Many Dimensions") when he criticizes Helmholtz's "romances of Flatland."[23]

When not treated as a joke, *Flatland* has tended to be approached in ways that divorce it from its cultural position in the debate over non-Euclideanism and its implications. Historically, literary critics have treated it as an early example of science fiction and fantasy, while scientists and mathematicians have used it as a clever way to introduce their students to concepts of dimensionality and non-Euclideanism. It has only been recently that the book has been brought back toward the center of the study of Victorian culture, and it will be to further that movement that I approach the novel here.[24]

## ABBOTT AND BACON

Edwin Abbott attended the City of London School and St. John's College, Cambridge, where he was senior classic and senior chancellor's medalist in 1861.[25] In 1865, at the age of twenty-six, he became headmaster of the City of London School, remaining there until 1889. During his tenure he developed a reputation as one of Britain's leading educators, instituting numerous curricular reforms, including the introduction of chemistry and English literature as compulsory subjects. He also produced several books on teaching, from textbooks on composition to a widely used Shakespearean grammar to a work on Latin prose. Yet at the same time he also wrote three books on Bacon, several works on theology that included historical/theological novels set just after the time of Christ, and *Flatland*. When he retired from teaching, it was to devote the remainder of his life to the development and publication of his theological system.

As Jann has noted, Abbott's intellectual positions were "characteristically—and sometimes controversially—liberal" (290). When it came to Bacon, Abbott did not shy away from a bitter public dispute with James Spedding, Britain's leading authority on Bacon and the primary editor of the magisterial edition of Bacon's *Works* that began appearing when Abbott was at Cambridge. The dispute began in 1876 with the publication of Abbott's edition of Bacon's *Essays* and continued until Spedding's death in the early 1880s.[26] Most of Abbott's work on Bacon, and certainly the bulk of his controversy with Spedding, concerned Bacon's character. Yet unlike both Spedding and Macaulay, Abbott saw Bacon the philosopher and Bacon the politician as similar rather than separate—his Machiavellian political pragmatism was in keeping with his urgings for an objective approach to the study of nature. He had that "cool, dispassionate, impartial way of looking at things which a man of science should have. . . . He was aware of the scientific danger of ignoring inconvenient facts and constructing convenient facts: and he had something of the scientific simplicity, taking things as they are and not as he would have liked them to be."[27]

This traditional Baconian view of the scientist's objectivity is paralleled initially by a fairly strong defense of Baconian induction and Bacon's philosophical importance. Of Macaulay's criticism that induction is something every person practices every day, Abbott claims that "the Essayist has scarcely done justice to the strictness and elaborateness of the Baconian Induction," that "[t]he Inductive process of Lord Macaulay's 'plain man' . . . is far below the level of the Baconian Induction" (lxxv). Not only does Abbott agree with Bacon that his induction is a new and important innovation, a distinct improvement over Aristotle's enumerative induction, he also supports Bacon's contention that the new induction is virtually foolproof. "If cautiously and scientifically used," writes Abbott, Baconian induction "cannot lead wrong" (lxxviii). Indeed, Abbott specifically refutes Ellis and Spedding's assessment that Bacon's method is "practically useless." Although the "successful discoverer" is "the man of all men most likely to see in the New Induction but a mere paper-philosophy" because he thinks "his discoveries have never been made in that way," this discoverer "has used it, or has used his abridgement of it, without knowing it. . . . [He] who so ungratefully decries Bacon's system is really claimed . . . as an adherent, as one of those unconscious pupils . . ." (lxxxix–xc).

Such enthusiastic endorsement of the specific details of Bacon's method, however, give way to the vague comments so typical in the latter half of the century about the value of the "spirit" of Bacon's work. The Method of Exclusions Abbott calls "the heel of our Achilles," denying both the assumption of the simplicity of nature on which it is based and the claims for absolute certainty subsequently based upon it (lxxxvi). Like Bacon the politician, Bacon the philosopher is careless about details, inconsistent, and unable to follow his own rules; the example of heat shows that although Bacon knew the value of "what we would call now-a-days a working hypothesis," he wrongly thought that this hypothesis must be obtained by applying the "Achilles heel" of exclusion (lxxxii), a mechanical process that leaves little room for the unfettered operation of the imagination. Tacitly acknowledging the uselessness of the "elaborate technicalities" and "complicated machinery" of Bacon's method, Abbott is forced to retreat to more general words of praise: "Probably Bacon has done much to raise the general level of scientific thought; and in this general rise the great scientific discoverers have, though unconsciously, shared. . . . The standard of science throughout the world has been raised by the *Novum Organum*. Put aside the details of its complicated machinery as useless, yet the spirit of it must be confessed to diffuse in all readers the love of Truth, and the sense of Law; and these to make up the very atmosphere of Science" (xc–xci).

And yet, in a final twist, Abbott raises this general praise to a new level, to the claims that Bacon is virtually responsible for the enabling scientific assumption that nature is ordered and humans can discover that order, and that nature is a second scripture the study of which leads to God. Having earlier endorsed the Bacon-as-Moses analogy—Bacon is "the Prophet of the New Logic" (xxxvii) who gives us a view of the fertile "Promised Land of Science" (xxiv)—Abbott exchanges Moses for John the Baptist in language that interweaves the words of Bacon with those of the New Testament in an effort to make new claims for the value of Bacon's "spirit." The "largest debt we owe to Bacon" is that

> No man who has ever been touched by the spirit of the *Novum Organum* can easily relapse into the belief that the world is a collection of accidents, or that its ways are past finding out. To have imbued and permeated mankind with a sense of the divine order and oneness of the Universe and of its adaptation to the human mind; to have turned men's thoughts to science as a divine pursuit, sanctioned by Him who hath *set the world in the heart of man*, and worthy to be called the study of the second scripture of God; to have proclaimed in undying words that all men shall learn that volume of God's works if they will but condescend to spell before they read; that all may be admitted into the Kingdom of Man over Nature by becoming as little children, and by learning to obey Nature that they may command her, and to understand her language that they may compel her to speak it—this Gospel to have proclaimed, and thus to have prepared the way for the scientific redemption of mankind, entitles Bacon to claim . . . that he wrote about science . . . like a Priest, like a Prophet of Science. (xclv–xcv)

Nine years later, however, after the debates with Spedding and the publication of *Flatland* Abbott's view of Bacon's philosophy is much more in line with Spedding's. In *Francis Bacon: An Account of His Life and Works*, Abbott criticizes Bacon's view of the orderliness of nature, the same view he earlier called the largest debt we owe to Bacon and to the *Novum Organum*. According to Abbott, Bacon believed too readily that the removal of Aristotelian mistakes about induction would lead directly to easy discovery: "he conceived a sanguine prejudice not only that Nature was orderly, but that her secrets must be readily discovered in an orderly and almost mechanical manner."[28] Following Ellis and Spedding and invoking the term Tyndall had made famous, Abbott argued that Bacon's lack of familiarity with contemporary science kept him unaware "of the indefinable combination of scientific imagination with scientific toil and observation, necessary to constitute an original Discoverer—[Bacon] early persuaded himself into the belief that it would be a matter of no great difficulty for him to elaborate such an Art of Induction as would make the discovery of Nature's secrets little more than a mechanical routine" (345–46). Rather than seeing today's discoverers as unconsciously following Bacon's methods in spite of their denials, Abbott now regards Bacon's methods as inapplicable—both then and now—because of their "undue neglect of the use of the Imagination in scientific research" (409).

Here as in the earlier work, Abbott falls back on the general spirit of Bacon's influence, but this time Abbott's sense of that influence is much more muted. Bacon's is "a vague, indefinite kind of influence, by no means such as Bacon himself would have preferred" (414). Yet this vague and indefinite influence, not the specific elements of his inductive philosophy, is the reason why Bacon is still read, and the reason, Abbott believes, why his influence will endure.

The changes in Abbott's assessment of Bacon's system are important as an indication of the broad cultural consensus about Bacon which developed in the second half of the century, but more important is the fact that this change occurs as a result of Abbott's recognition of the importance of the imagination in the acquisition of knowledge, whether scientific, religious, or moral. As Jann has argued, "the idea of imagination working through appearances to higher truth is for Abbott the fundamental mechanism of both science and religious thought" (294). We can see this clearly in *The Kernel and the Husk* (1887), where Abbott says he was led to the conclusion that

> a great part of what we call knowledge does not come to us, as we falsely suppose it does, through mere logic or Reason, nor through unaided experience, but through the emotions and the Imagination, tested by Reason and experience. Even in the world of science . . . the so-called "laws and properties of

> matter," nay, the very existence of matter, were nothing more than suggestions of the scientific Imagination aided by experience. A great part of the environment and development of mankind appeared to have been directed towards the building-up of the imaginative faculty, without which, it seemed that religion, as well as poetry, would have been non-existent.[29]

For Abbott, however much we depend on experience, no amount of experience can justify our generalizations and predictions; the assumption of the uniformity of nature or of cause and effect is an act of *faith* in which our Imagination plays a crucial role. "The very names 'cause' and 'effect' are due not to observation, nor to demonstration, but to faith. The name and the notion of a Law of Nature, are nothing but convenient ideas of the Scientific Imagination, based upon faith" (75).

Jann is right that Abbott is trying to restore a spiritual element to the materialist science of Tyndall and Huxley (297). Yet it is equally important to see that Abbott appropriates the "Scientific Imagination" from Tyndall and that, as we saw in Chapter 1, Tyndall's language itself endeavors to maintain a sense of mystery about the activities of that imagination. Whether Abbott sees himself as subverting Tyndall or not, he rightly identifies and exploits that element of Tyndall's thought most congenial to his own philosophy, what Jann terms Abbott's "rebuke of both materialist science and fundamentalist religion" (294). But an even more important influence on Abbott's epistemological views would appear to be Coleridge's version of German transcendentalism, appropriated by Abbott for his own purposes.

Abbott opens his discussion of Reason, Understanding, and Imagination in *The Kernel and the Husk* by noting that he uses Reason "in a sense for which Coleridge (I believe) preferred to use 'Understanding'" (47n). The "I believe" is a confession, even as Abbott acknowledges his debt to Coleridge, both of no deep knowledge of Coleridge's system and of a willingness—immediately put into practice—to play with Coleridge's terms. Thus Abbott explains that his use of Reason corresponds to the "popular sense" of Reason as the "mathematical, logical, and ordinary processes of arguing," and this is why he says it corresponds to the Coleridgean Understanding. Since Coleridge defined logic as the science of the Understanding, this description is basically accurate, but Abbott's use of the popular meaning of the term Reason would have troubled Coleridge not simply because he found such meanings imprecise, but because "the business of common logic," however "indispensable in all the moral and forensic concerns of life," is "utter[ly] inadequate in the investigations of nature, and . . . in those of the pure reason."[30] Abbott's definition of the Imagination, at least in relation to

Coleridge, is similarly imprecise. On the one hand, Abbott says the Imagination might be called the "higher Reason," and it does have something in common with Coleridge's Reason. Abbott's Imagination is more important than both the Senses and the Reason, for most of our knowledge comes through neither the unaided sensations nor the unaided reason, but through the Imagination. On the other hand, Abbott's Imagination also seems to serve the mediating function that Coleridge's Imagination serves. In Abbott's scheme, the Imagination transforms and recombines the objects of the Senses, and the Reason, the logical faculty, with the aid of the experience built up from the Senses, then tests the validity of these products of the Imagination. "The Imagination," says Abbott, "is the 'imaging' faculty of the mind. It does not, strictly speaking, create, any more than an artist, strictly speaking, creates. But as an artist combines lines, colours, shades, sounds, and thoughts, each one of which by itself is familiar to everybody, in such new combinations as to produce effects that impress us as original and unprecedented, so does the Imagination out of old fragments make new existences and unities" (47). In keeping with Abbott's view that we are led to truth through illusion, the Imagination "rarely conducts us to truth without first leading us through error. Its business is to find likenesses and connections and to suggest explanations, not to point out differences, and make distinctions, and test explanations; these latter tasks are to be accomplished not by the Imagination but by Reason with the aid of enlarged experience. . . . [I]f the Imagination did not first suggest the ideas on which the Reason is to operate, we should never have anything worth calling knowledge" (49–50).

Although Coleridge would have probably been dismayed by many of Abbott's revisions, it is easy to see why Abbott found Coleridge congenial: both men developed a system of knowledge that sought to avoid the extremes of materialism and idealism and that brought together religion, science, and art into a unified whole. If Abbott's discussions of methodology are less subtle than Coleridge's, that lack of subtlety makes the connection Abbott sees between science and art even more obvious. For Abbott, the scientific and the artistic imaginations, though apparently at odds, are "really not dissimilar" (54), and the association of the imagination with the production of hypotheses in scientific method is direct. Protesting against what he calls "the popular fallacy that the Imagination is an abnormal faculty, limited to poets and painters and 'artists,' mostly illusive, and always to be subordinated in the search after truth," Abbott argues that the Imagination "lies at the basis of all knowledge; that it is no less necessary for science, for morals, and for religion, than for artistic success; and that the illusions of Imagination are the stepping-stones of Truths" (54). And, significantly, the Imagination is also at work in the two methods of the Reason, in the processes of induction and deduction. Induction is more dependent on the Imagination than deduction, for while deduction is primarily negative and eliminative, induction is positive and speculative. "[S]urely you must admit," Abbott writes, "that the initial part of the task [of induction] falls not upon the Reason but upon the Imagination; which sees likenesses and leaps to general conclusions, mostly premature or false, but all containing a truth from which the falsehood must be eliminated" (55). What Abbott is endorsing here is very un-Baconian. Bacon rejected such "leaps to general conclusions" precisely *because* they were "mostly premature or false," and he saw elimination as the essential feature of induction, occurring before rather than after any general conclusion. For Abbott, however, "Imagination is the mother of working-hypotheses" (50): we first imagine something to be true and then allow the Reason to test whether it is true.

## ABBOTT AND EUCLID

Determining Abbott's views about non-Euclidean geometry, and relating those views to his position on scientific method and the scientific imagination, is more difficult. But we can begin with some of the things Abbott has to say about Euclidean geometry.

Abbott himself was no friend of Euclid. He recalled his own schoolboy experiences with the *Elements* as unpleasant ones.[31] Such experiences were not uncommon. Even a brilliant mathematician like Sylvester admitted that the "early study of Euclid made me a hater of geometry" and urged that education would be far more effective if "Euclid [were] honourably shelved or buried 'deeper than did ever plummet sound' out of the schoolboy's reach."[32] Not surprisingly, in this same period that non-Euclidean geometry was being developed, an anti-Euclidean movement formed, its goal to "shelve" Euclid as a school textbook.[33] The same figures were often members of both groups, and the same problem that initiated the construction of alternative geometries—Euclid's parallel postulate—was at the center of the construction of alternative textbooks. Abbott and Clifford were both members of the Association for the Improvement of Geometrical Teaching, which produced its own syllabus for the teaching of geometry, but other textbooks flooded the market, including one by Francis Cuthbertson, the mathematical master under Abbott at the City of London School.

If Abbott, like Sylvester, was more interested in shelving Euclid than in defending him, he also shared with Sylvester the view that mathematics was not some sterile, unimaginative, deductive science, and this view seems to have developed with his changing assessment of Bacon. The notion that mathematical truth relies solely on "pure Reason" and not at all on Imagination Abbott termed "grotesque" in *The Kernel and the Husk* (29). Yet mathematical truth is not less truth for requiring the Imagination:

> The whole of what we call "Euclid" is based upon a most aerial effort of the Imagination. We have to imagine lines without thickness, straightness that does not deviate the billionth part of an inch from perfect evenness, perfectly symmetrical circles, and—climax of audacity!—points that have "no parts and no magnitude"! Obviously these things have no existence except in the dreams of Imagination; yet Euclid's severe reasoning applies to none but these things. If you step from your ideal triangle in Dreamland into your material triangle in Chalk-land, you step from absolute truth into statements that are not absolutely true. . . . In a word the whole of Geometry is an appeal to the Imagination in which the geometer says to us, "I know that my propositions are not exactly true except with respect to invisible, ideal, and imaginative figures, planes, and solids. These ideas, therefore, you must endeavour to imagine. In order to relieve the strain on your imagination, I will place before you material and visible figures about which my reasoning will be approximately true. From these I must ask you to try to rise upward to the imagination of their archetypes, the immaterial realities." (30–31)

Characteristically, Abbott negotiates between empiricist and idealist conceptions of geometry. If geometrical figures are idealizations of real points and lines and planes, Abbott's emphasis, like Cayley's, is more on these "archetypes" than on the "material, visible" figures. Although he defends mathematics as an inductive science, Abbott sees these inductions ultimately leading us to a realm of faith. Because Euclidean geometry itself depends on imagination, it is no necessary truth except in the limited sphere of archetypes and severe reasoning, yet for Abbott the Dreamland triangle is an apparently higher *reality*. This Platonic view is justified, however, in a way both Plato and Coleridge would no doubt have been troubled by: "that [these imaginary figures] are realities, and that our conclusions about them are real and true, is proved by the one test of truth: our conclusions *work*. . . . A perfect circle you never saw and never will see; yet it is as real as a beefsteak and a pint of porter. I believe in a perfect circle by Faith; I accept it with reverence as an impression . . . on the Mind of the Universe, which He has communicated to me. What is more, I believe that He intended us to study this and other immaterial realities that our minds might approximate to His" (32).

Yet Abbott was uneasy with the specific lengths to which other writers could take the connection between the fourth dimension and the spiritual world. Russell casually associates Abbott with "a few spiritualists" in the use of the Flatland-Sphereland relation to imply the existence of the fourth dimension, but if the specific identification is wrong, it is right to say that *Flatland* was appropriated by those who saw the fourth dimension as a way of accounting for spiritual, and sometimes specifically Christian, phenomena.[34] A. T. Schofield's *Another World, or The Fourth Dimension* (1888) makes such extensive use of *Flatland* that when Schofield turns from an extended summary of Abbott's book to an investigation of the "powers and laws" of the fourth-dimensional "spiritual world" and its "correspondence to the spiritual claims of the Christian religion," he claims to be carrying *Flatland*'s "line of argument . . . to what seems to me its true and necessary conclusion."[35] Less overtly Christian were the many works, some fictional and some not, of C. H. Hinton, who wanted to assist readers in visualizing the fourth dimension and to connect higher dimensions with a spiritual world.[36]

In *The Kernel and the Husk*, Abbott expresses both pleasure and concern—though indirectly, for his authorship of *Flatland* was still unacknowledged—about this appropriation of his work. He admits that he is attracted to the "harmless fancy" that "there are two worlds; one visible, terrestrial, and material, the other invisible, celestial, and spiritual; and that whatsoever takes place down here takes place first (or simultaneously but causatively) up there." A similar fancy—which he tells the reader he would know "if you would read a little book recently published called Flatland, and still better, if you would study a very able and original work by Mr. C. H. Hinton"—is that

> a being of Four Dimensions, if such there were, could come into our closed rooms without opening door or window, nay, could even penetrate into, and inhabit, our bodies; that he could simultaneously see the insides of all things and the interior of the whole earth thrown open to his vision; he would also have the power of making himself visible and invisible at pleasure; and could address words to us from an invisible position outside us, or inside our own person. Why then might not spirits be beings of the Fourth Dimension? Well, I will tell you why. Although we cannot hope ever to comprehend what a spirit is . . . yet St. Paul teaches us that the deep things of the spirit are in some degree made known to us by our own spirits. Now when does the spirit seem most active in us? . . . Is it not when we are exercising those virtues which, as St. Paul

says, "abide"—I mean faith, hope and love? Now there is obviously no connection between these virtues and the Fourth Dimension. Even if we could conceive of space of Four Dimensions—which we cannot do, although we can perhaps describe what some of its phenomena would be if it existed—we should not be a whit the better morally or spiritually. (258–59)

Although Abbott is interested in what the phenomena of the fourth dimension might look like, and although he clearly sees *Flatland* as suggesting some of these phenomena (the ability of the Sphere, for example, to "make himself visible and invisible at pleasure" to A Square, and to "address words" to him from an "invisible position"), he is uncomfortable with the specific identification of the fourth dimension with the spiritual world. Although we can describe what some of the fourth dimension's phenomena might be like, we still cannot, as Cayley insisted, conceive of four-dimensional space. Twenty years later, Abbott's position on faith in this higher reality, as well as the language used to express it, remains the same. In a discussion of miracles and visions, Abbott argues that only some are literally true, although some are "ordained by God—in accordance with the laws of the spiritual world of which we have very little knowledge although it both surrounds us and is in us—to give us glimpses of truth beyond the truths of what we call 'matter,' and to be more real than any of what we call 'material' occurrences. As Solidland may be called more 'real' than Flatland, so Thoughtland may be found more real than Factland. We really *know* nothing whatever about what is 'real.' We have only faith and feeling about it."[37] Here, too, Abbott is trying to find an acceptable middle ground. He is willing to go a long way in invoking Tyndall's scientific imagination or in echoing the views of those who asserted that matter is a convenient construct. In Abbott's words, "my private fancy is, that there is no such thing as matter (though of course physicists may assume it, for a working-hypothesis, as Euclid assumes non-existent and non-possible points and straight lines). . . . " But "[s]peculations about the ultimate material basis of phenomena appear to me barren, and not to be indulged in except as amusements of an idle hour, like excursions into the Fourth Dimension."[38] If materialists are unjustified in claiming that all phenomena are ultimately material phenomena, spiritualists and fundamentalists are similarly unjustified in claiming that they must be spiritual. These references pointing back to *Flatland* are no accident here, for Abbott specifically recalls his preface to the second edition of the book, a preface in which he reports A Square as hoping that, "taken as a whole, his work may prove suggestive as well as amusing, to those Spacelanders who—speaking of that which is of the highest importance, but lies beyond experience—decline to say on the one hand, 'This can never be,' and on the other hand, 'It must needs be precisely thus, and we know all about it.'"[39]

## THE GEOMETRY OF *FLATLAND*

Abbott's willingness to treat Euclidean geometry not as an example of necessary truth but as an example of the need for faith and the imagination suggests some sympathy with non-Euclidean geometry and the philosophical implications drawn from it. The momentum of the basic argument of *Flatland*, moving from the second dimension to the third and even beyond, also suggests suspicion of those who accept non-Euclidean geometries but deny non-Euclidean spaces. But Abbott's view that faith and imagination lead to this world of higher reality, a spiritual world, a Thoughtland, implies that there is a limit to his support for empiricist efforts to limit geometry to a study of our experience of space.

These concerns begin to emerge in A Square's dedication, offered "In the Hope that / even as he was Initiated into the Mysteries / Of THREE DIMENSIONS / Having been previously conversant / With ONLY TWO / So the Citizens of that Celestial Region / May aspire yet higher and higher / To the Secrets of FOUR FIVE OR EVEN SIX Dimensions / Thereby contributing / To the Enlargement of THE IMAGINATION / And the possible Development / Of that most rare and excellent Gift of MODESTY / Among the Superior Races / Of SOLID HUMANITY." The book's satiric element prevents us from taking anything or its opposite too literally, yet this dedication immediately invites us to employ and enlarge our imagination, that vital pathway to truth in Abbott's thought, in the matter of higher dimensions. To assume that higher dimensions do not and cannot exist suggests a lack of modesty—it confirms, in A Square's words, that we are "slaves" of our "Dimensional prejudices" (xii). The book itself dramatizes these prejudices by having the representative of each successive level of dimensionality—including the Sphere—deny that a higher level exists. Although no effort is made to show us the fourth dimension, A Square, carrying through "the Argument from Analogy of Figures" that the Sphere used in explaining the third dimension, describes some of the characteristics of a four-dimensional figure (93–96). Carried away by enthusiasm, A Square initially asserts that "of a surety there is a fourth dimension" (94), but he quickly retracts that statement, for he realizes that analogical arguments cannot provide certainty: "I cast myself in faith upon conjecture, not knowing the facts" (95). Here again, scientific and religious conclusions are dependent on faith and imagination. Although A Square has visited Spaceland, he "cannot . . . comprehend it, nor realize it by the sense of sight or by any process of reason; I can but apprehend it by faith" (xii). The senses and the pure reason, characteristically in Abbott, are inadequate by themselves.

The ability to visualize, whether with the body's or the mind's eye, was crucial to Helmholtz's Flatland narratives. Abbott's extensive discussions of what his Flatlanders see, how they experience their space, are an elaborate expansion of Helmholtz's initial example. The logic of the "Argument from Analogy" and the willingness to discuss the geometrical configuration of a four-dimensional figure further indicate that Abbott is receptive to the possibility of the existence of non-Euclidean worlds. But if "dimensional prejudices" are exposed, the empiricists' claims that they are facilitating the ability to conceive such worlds are viewed with suspicion. The "process of reason" that characters in Flatland use for discussing higher dimensions does not enable them to "comprehend" those dimensions. When A Square is taken into Spaceland, his enthusiasm for analogical arguments carries him even beyond the fourth dimension: "In that blessed region of Four Dimensions, shall we linger on the threshold of the Fifth, and not enter therein? Ah, no! Let us rather resolve that our ambition shall soar with our corporeal ascent. Then, yielding to our intellectual onset, the gates of the Sixth Dimension shall fly open; after that a Seventh, and then an Eighth—" (96). Yet in his reminiscence A Square admits that he was getting carried away. "Nothing could stem the flood of my ecstatic aspirations," he recalls. "I was intoxicated with the recent draughts of Truth" (96).

The Sphere is initially critical of A Square's argument for higher dimensions and angrily insists that space is three-dimensional. Indeed, when A Square presses him for a glimpse of the fourth dimension, the Sphere responds that "the very idea of it is utterly inconceivable" (93). The Sphere is of course wrong—but only that the *idea* is inconceivable, not that four-dimensional space is inconceivable—and after hurling A Square back into Flatland, he admits that he has no basis for his own dimensional prejudice. The Sphere urges A Square to spread "the Gospel of Three Dimensions" and "initiates" him into even "higher mysteries" of "Extra-Solids" and "Double Extra-Solids" (100). The religious language underscores the view that the mysteries of geometry and of the New Testament both require faith and imagination. For the reader this is doubly true, since the vision and its interpretation are revelation within revelation. Their absolute truth cannot be confirmed—and of course the other Flatlanders reject A Square's story and imprison him when he attempts to preach his new Gospel—but can only be accepted by faith. The Argument from Analogy, which proceeds via a series of imaginative constructs, can only suggest or direct, not prove or disprove.[40]

This criticism of analogy is also evident in Abbott's use of natural theology's argument from design, which tended to make the status quo a naturally ordained—and thus ultimately a divinely ordained—state of affairs. A Square often invokes the language of natural theology to naturalize the specific social and political ideologies of Flatland that Abbott is critiquing. For example, rank in Flatland is determined by the number of sides a figure has: women are line segments, workers and soldiers are isosceles triangles whose acute angles increase from generation to generation until they become equilateral, after which descendents can successively add a side and become squares, pentagons, hexagons, etc. (Hence A Square's sons are Pentagons and his grandson is a Hexagon.) Women and Isosceles, because of their sharp points and large numbers, are potentially dangerous to Flatland's ruling classes, but they lack the brain power to take advantage of it, a fact that A Square explains thus:

> a wise ordinance of Nature has decreed that, in proportion as the working-classes increase in intelligence, knowledge, and all virtue, in that same proportion their acute angle (which makes them physically terrible) shall increase also and approximate to the comparatively harmless angle of the Equilateral Triangle. Thus, in the most brutal and formidable of the soldier class—creatures almost on a level with women in their lack of intelligence—it is found that, as they wax in the mental ability necessary to employ their tremendous penetrating power to advantage, so do they wane in the power of penetration itself.
>
> How admirable is this Law of Compensation! And how perfect a proof of the natural fitness and, I may almost say, the divine origin of the aristocratic constitution of the States in Flatland! By a judicious use of this Law of Nature, the Polygons and Circles are almost always able to stifle sedition in its very cradle. (11)

The ruling classes in Flatland use this "Law of Nature"—which in fact has numerous exceptions and which is constantly being tampered with rather than submitted to in an effort to get around another law, that with increasing sides comes decreasing fertility—to legitimize and maintain their control by making that control natural and scientifically ordained.

The suppression of women and irregular figures is also justified by such arguments. The fact that women remain lines rather than increasing their angles through generations suggests to A Square that the expression "Once a Woman, always a Woman" is a "Decree of Nature" (17). Men are the natural masters of women, who, because they have no angle, are "wholly devoid of brain-power" (15). Yet we can admire the "wise Prearrangement" which has "ordained" that if women have nothing to look forward to, they do not have the brain power to look forward anyway, and their lack of memory prevents them from recalling "the miseries and humiliations which are . . . a necessity of their existence. . . ." Like women, Irregular figures also live miserable and humiliating lives and are seen as morally and physically dangerous. In many parts of Flatland they are destroyed at birth. According to A Square, "the whole of the social life in Flatland"—indeed, of civilization itself—"rests on the fundamental fact that Nature wills all Figures to have their sides equal" (30). Thus it is "an axiom of policy" that "Irregularity is incompatible with the safety of the State" (31–32). But this "axiom," like the other "laws" and "decrees" of nature, is not a divinely ordained necessary truth but a culturally constructed generalization. A Square acknowledges that there are those who deny any "necessary connection between geometrical and moral Irregularity" (31) and who argue that the treatment of Irregulars is what causes their antisocial behavior, but he rejects these contentions out of hand: "I . . . have never known an Irregular who was not also what Nature evidently intended him to be—a hypocrite, a misanthropist, and . . . a perpetrator of all manner of mischief" (32). In A Square's natural theological world, there is no place for anything irregular, anything that does not fit the design of Nature.

Knowledge of how many sides another Figure has, and whether or not those sides are regular, is vital in Flatland, yet Flatlanders cannot know these things directly because what they see are not sides but lines. One method for determining numbers of sides is through feeling the other figure, but this method is considered (by the upper classes) not only inefficient, but primitive and rude. The upper classes therefore practice and teach to their children the art of sight recognition, which, as Jann notes, requires both induction and inference, but which is also inseparable from a sophisticated understanding of geometry (302). Since the sides of Flatlanders glow and the atmosphere in Flatland is foggy, it is possible to determine the number of sides and size of angles of another figure by attending to the brightness of the line presented. Sight recognition is used by the "Polygonal Class" to reinforce class distinctions, which are explicitly said by A Square to reflect intellectual distinctions. Yet the ability to recognize by sight is in fact an acquired skill, requiring much education and experience. Sight recognition is an "unattainable luxury" among the lower classes because even "[a] common Tradesman cannot afford to let his son spend a third of his life in abstract studies" (28). Polygonal children, on the other hand, are sent to "the illustrious University of Wentbridge," where sight recognition is taught by "the Learned Professors of Geometry" (26).[41] Women lack the ability even to count, but that is because they are not educated at all, a fact A Square regrets only because of its enfeebling effect on the intellects of *male* figures, who are forced into a "bi-mental" existence by the need to talk down to the mental level of women (52–53).

Anything that threatens to undermine the importance of sight recognition invites, in A Square's words, "a total destruction of all Aristocratic Legislatures and . . . the subversion of our Privileged Classes" (40). At one point in Flatland's history, the discovery of colors and the development of painting led figures to decorate themselves both for aesthetic purposes and for easier identification. But this in turn virtually eliminated the need for sight recognition, paving the way for the lower classes to assert that there was no great difference between themselves and the polygons, and that "all individuals and all classes should be recognized as absolutely equal and entitled to equal rights" (37). The lower classes first argued that "Distinction of sides is intended by Nature to imply distinction of colours" (34), but then they turned the tables on the aristocracy, contending that "distinction of colours" was "a second Nature" more appropriate than sides and sight recognition and geometry for distinguishing one figure from another. They lobbied for a "Universal Colour Bill," but this "Chromatic Sedition," A Square informs us, was crushed. The reactionary handling of this "sedition"—a handling that A Square both endorses and admires—is impossible to miss. The Circles pretend to accept the Colour Bill but in Marc Antony-like fashion are able to set the lower classes against their own leaders, who are brutally killed. Color is abolished, those figures suspected of irregularity are executed without being measured, and "every town, village, and hamlet was systematically purged of that excess of the lower orders . . . brought about . . . by the violation of the . . . natural Laws of the Constitution of Flatland" (44).

Putting aside the obvious importance of this episode for contemporary British social politics, Abbott here is also using A Square's reactionary stance to expose the fallacy in arguments that, in the fashion of natural theology, naturalize the artificial. Distinction based on color is clearly artificial, but so too is distinction based on sides, which has more to do with access to education than with the size of a figure's brain-angle. When A Square dismisses as a "so-called axiom" (35) the innovators' claim that distinction of sides implies distinction of color, we are able to see that his social axioms, like the geometric ones on which they are based, are also arbitrary and self-serving assumptions.

The "Colour Revolt" was part of Flatland's richest period of artistic creativity, and even A Square regrets that life in Flatland is now so dull. Yet he sees that this period was also dangerously anti-intellectual and antiscientific. What he cannot seem to see is that if the suppression of the Colour Revolt protected Flatland from relying too much on the senses and not enough on the intellect (the difference between color recognition and sight recognition), it has resulted in the opposite imbalance, with too much emphasis on the unaided Reason, especially among the upper-class males. This is evident in the attitude toward the women, who are considered "deficient in Reason but abundant in Emotion" and are associated with such "irrational and emotional conceptions" as love, duty, right, wrong, pity, and hope (52). When male figures are three years old, they are "taught to unlearn the old language" of "Mothers and Nurses" and replace it with "the vocabulary and idiom of science" (53), a language which transmutes irrational emotions like love and duty into "the anticipation of benefits" and "necessity or fitness" (52). Yet in Abbott's view, the vocabulary and idiom of science cannot be shaped by reason alone—it requires emotion, imagination, faith. The implication in *Flatland*, as in Abbott's nonfictional works, is that the scientific and the imaginative must be brought together. The senses and the reason are incomplete by themselves. As with Coleridge's view of the imagination as an "esemplastic" power, for Abbott it is the imagination that "out of old fragments makes new existences and unities."

*Flatland* also takes up this question of balance in the issue of non-Euclidean geometries versus non-Euclidean spaces. Because A Square's precocious young grandson is doing so well at his practical mathematical lessons, A Square decides to "reward" him "by giving him a few hints on Arithmetic, as applied to Geometry." A Square's "hint" is that it is possible to determine the area of the inside of a square (something the two-dimensional Flatlanders cannot see directly) by squaring the length of a side. The Hexagon's unnerving reply is that if the arithmetical operation of raising a number to a power can be represented geometrically, then "$3^3$ must mean something in Geometry." But A Square tells him that $3^3$ has no meaning, "not at least in Geometry, for

Geometry has only Two Dimensions." Unconvinced, the Hexagon persists: "Well, then, if a Point by moving three inches, makes a Line of three inches represented by 3; and if a straight Line of three inches, moving parallel to itself, makes a Square of three inches every way, represented by $3^2$; it must be that a Square of three inches every way, moving somehow parallel to itself (but I don't see how) must make Something else (but I don't see what) of three inches every way—and this must be represented by $3^3$" (69–70). The flustered A Square sends his grandson to bed with admonitions about not talking nonsense, but it is only a few moments later that the Sphere announces his presence with the words, "'The boy is not a fool; and $3^3$ has an obvious geometrical meaning'" (70).

By denying that the third dimension has any geometrical significance, and by insisting that geometry has only two dimensions, A Square means that the third dimension cannot be represented spatially by two-dimensional beings in a two-dimensional space. The issues, as in the debate over Helmholtz's claims for his Flatland analogies, are whether it is possible to conceive (or imagine, or represent spatially—the terms are used differently by different people, but the common denominator is the ability to visualize) non-Euclidean and *n*-dimensional spaces, and whether the ability to construct non-Euclidean and *n*-dimensional geometries has any bearing on the existence of non-Euclidean and *n*-dimensional spaces. It is a crucial distinction even in *Flatland*, for throughout the book it is impossible for figures to conceive of a higher dimension than their own even if they can develop the properties of such a dimension analytically. The Sphere convinces A Square of the existence of the third dimension only by lifting him out of Flatland, and A Square's recollections of the third dimension grow dimmer when he is later returned to his two-dimensional world. But what is being satirized is not A Square's inability to move from the arithmetic concept of $3^3$ to its geometrical meaning but his refusal to accept even the *possibility* that $3^3$ has a geometrical meaning. Since Abbott's surface-dwellers live on a flat surface rather than a curved one, there is nothing in their experience of physical space to suggest a third dimension. The same is true for the Sphere and the fourth dimension, the King of Lineland and the second dimension, the King of Pointland and any dimension. The "obvious Geometrical meaning" of $3^3$ is obvious only to the Sphere, and the book's satire is equally directed against the Sphere for thinking that the meaning should be obvious to A Square.[42]

Figures in *Flatland* have no trouble following the *mathematical* aspect of the argument from analogy—they are perfectly able to develop different geometries. In this sense, there can be no doubt that Abbott endorses the mathematical validity of non-Euclidean geometry. But as we have seen, what was really at stake for mathematicians and philosophers was the existence of non-Euclidean spaces. On this question, Abbott reveals himself, in the act of protecting a middle ground, to be more conservative than is usually thought. Since for Abbott analogical arguments are suspect, the question is always what you are trying to make your analogy prove. In the case of *Flatland*, Abbott seems to be suggesting that the existence of different spaces is possible, but that possibility does not mean such spaces do exist (as Hinton and the spiritualists tended to claim), nor does it mean that we can perceive what experience in these spaces would be like (as Helmholtz and Clifford and Sylvester contended).

Yet in coming down between "this can never be" and "it must be precisely thus," Abbott is closer to Cayley than to the empiricist non-Euclideans. For both men, there is a distinction between the development of non-Euclidean geometries and the existence of non-Euclidean spaces, and between the ability to describe such spaces analogically and the ability to perceive them. Both treat Euclidean geometry from an idealist perspective: for Cayley geometry is the study of "purely imaginary objects" which are "the only realities"; for Abbott it is the study of "immaterial realities." Both, because of our inability to visualize non-Euclidean spaces, are skeptical about these spaces being part of any reality, immaterial or physical. The difference between them lies in Abbott's greater willingness to entertain the possibility of such spaces, in his recognition that if Euclidean geometry shows the necessity in science for the play of the imagination, the action of faith, non-Euclidean geometry requires even more faith. But even so, faith cannot extend to the unwarrantable "it must be precisely thus."

I offer these affinities between *Flatland* and Cayley's address not to suggest that Abbott is responding directly to Cayley, but to show that both Abbott and the mathematicians were concerned with the wider implications of shelving Euclid. Without the popularizations of Helmholtz and Clifford, and without the responses those popularizations generated, *Flatland* probably would not have been written. But Abbott's book does not simply use contemporary interest in geometry as the basis for a clever story. Rather, its satire participates as fully in Britain's reaction to non-Euclideanism as it does in Britain's social and political environment. That Abbott's views on geometry are incorporated into his broader epistemological views on religion, science, and art is more than not surprising: it is virtually unavoidable, given that all participants in the discussions of Euclid's axioms were well aware of, to use Helmholtz's words, the "immediate connections with the highest points regarding the nature of the human understanding."

Yet non-Euclidean geometry and Abbott seem to present at least two puzzling points in relation to scientific method. To defend mathematics as an inductive science, as both Sylvester and Abbott did, appears on the surface not to be in keeping with the movement of the debate on method discussed in Chapter 1. But if we keep in mind the substance of those defenses, and the nature of Huxley's dismissal of mathematics described at the start of this chapter, we find that non-Euclidean geometry plays a key role in bringing mathematics within the rubric of this changing conception of scientific method. For both Sylvester and Abbott, mathematics is inductive not in the traditional Baconian way, but in the manner that emerged during the nineteenth century, a manner that requires imagination, hypothesis, speculation. The axioms of Euclidean geometry are no longer the necessary assumptions of a dry deductive system but the working hypotheses used to test and describe our experience of physical space. Different sets of assumptions produce different geometrical systems, new mathematical worlds with possible implications for our own world.

The new non-Euclidean worlds established that even geometry requires the imaginative play so important to the physical and historical sciences. Yet in what may strike us as a rather odd reversal, the empiricist interpretation of the non-Euclidean revolution was contained, at least initially, by British Kantians acting as a conservative force. While this success obviously attests to the growing influence of Kantian philosophy in Britain and to a breakdown in divisions of philosophy along national lines (it was hard to criticize "German" idealism with Helmholtz and Riemann as leading non-Euclideans, difficult to isolate "British" empiricism with Cayley invoking Plato and Kant in his critique of Helmholtz's views), the presence of enough scientists and mathematicians aligned with the Kantians in a conservative camp, opposing the powerful empiricist arguments of Sylvester, Clifford, and Helmholtz, requires an explanation. A possible answer, one that arises directly from the non-Euclidean debate and draws on the positions of Abbott, involves what these men were trying to conserve.

As Russell noted in 1897, geometry had long been the best proof that knowledge is not simply experiential, and Kant's explanation for the necessary and a priori truth of Euclid's axioms the most compelling.[43]

For Coleridge, following Kant and Shelling, "Geometry . . . supplies philosophy with the example of a primary intuition, from which every science that lays claim to evidence must take its commencement."[44] But if geometry and philosophy commence in a "primary intuition," the latter ends, in Coleridge's transcendentalism, in God. Thus Abbott, following in his turn Coleridge, argues that the imaginary figures of geometry are immaterial realities communicated to our minds by a God who wishes us to have access to the divine mind. Defending the Kantian conception of space from the attacks of non-Euclideans was thus the preservation of this important means of access to the divine, or at the very least to a source of knowledge beyond human experience. Whereas the nineteenth century had opened with frequent contrasts between the vagaries of Continental "speculation" and "generalization" and the cautious solidity of British empiricism, the non-Euclidean debate pitted against each other two sides that both acknowledged the value of the speculative imagination. Although Cayley was recognized as one of Britain's most brilliantly imaginative mathematicians, and although Abbott built a philosophical and theological system around the Imagination, both were ultimately conservative in their responses to non-Euclidean geometry, at least in relation to the likes of Helmholtz and Clifford and Sylvester—who had, after all, with their aggressive popularizing, set the terms of the debate. In his liberal reformist desire to shelve Euclid, Abbott nonetheless wanted to keep him within honorable reach.

# 7

# "The Methods Have Been of Interest"

## Sherlock Holmes, Scientific Detective

In a discussion of science in his collection of autobiographical literary essays, *Through the Magic Door* (1907), Arthur Conan Doyle writes that "[t]he mere suspicion of scientific thought or scientific method has a great charm in any branch of literature, however far it may be removed from actual research. Poe's tales, for example, owe much to this effect, though in his case it was pure illusion."[1] Having already acknowledged that any detective story writer must follow in the path set by Poe in the Dupin stories, Doyle essentially implies that his "little development" is the *real* introduction of scientific thought or scientific method into the detective genre.

This conclusion is reinforced in Doyle's autobiography, *Memories and Adventures* (1924). Recollecting the process by which he created Sherlock Holmes, Doyle briefly mentions the literary influence of Poe and Gaboriau before describing in greater detail the appearance and the diagnostic methods of his medical mentor at Edinburgh, Dr. Joseph Bell. "I thought of my old teacher Joe Bell, of his eagle face, of his curious ways, of his eerie trick of spotting details. If he were a detective he would surely reduce this fascinating but unorganized business to something nearer to an exact science." Like Holmes, Bell could deduce a patient's occupation and recent history from the careful observation of slight details: a mud stain on a trouser cuff, a callused thumb, a military bearing. "It is no wonder," Doyle continues, "that after the study of such a character I used and amplified his methods when in later life I tried to build up a scientific detective. . . ."[2] To highlight the methods of the "scientific detective," however, means to limit drastically the complexity of his character; Holmes, Doyle complains, "admits of no light or shade. He is a calculating machine, and anything you add to that simply weakens the effect."[3]

Doyle's characterization of Holmes as "a calculating machine" implies that method and personality are inseparable, though seemingly in a rather one-sided way. Generations of readers and critics have often described the man and his methods rather differently, balancing the calculating machine with an "artistic" personality and an intuitive method. Writing at almost the same time as Doyle, Willard Huntington Wright claimed that the intellectual appeal of detective fiction had caused the "plodding, hard working, routine investigator of the official police" to replace "the inspired, intuitive, brilliantly logical supersleuth of the late nineteenth century"—of which Holmes was the "foremost representative"—as the genre's typical hero.[4] Edmund Wilson, who enjoyed the Holmes stories but loathed almost all other detective fiction, called Holmes "intellect in the form of a romantic personality possessed by the scientific spirit."[5] The abstraction of Holmes into "intellect" and "scientific spirit" meshes nicely with Doyle's similarly dehumanizing conception of him as a calculating machine, while his "romantic personality" corresponds with Wright's sense of Holmes's inspiration and intuition.

Such comments helped to initiate a critical tradition in which Holmes's personality is said to consist of two sides, one "scientific" and the other "romantic." Holmes is a contradictory figure, now the scientist and now the artist, and his methods are equally contradictory. On the one hand, he is cold and impersonal, a single-minded ascetic with no interest in anything that doesn't relate to his profession; on the other hand, he is an intuitive genius, a bohemian artist who seeks refuge from boredom in cocaine, tobacco, and the violin. His detective methods seem to correspond to this dichotomy: on the one hand, he is an objective fact gatherer, magnifying glass in hand, reasoning from effect to cause without preconceived theories; on the other hand, he is an inspired spinner of theories who senses when someone is telling the truth and looks for the evidence that his hypotheses demand. For some critics, these sides are juxtaposed rather than fused because science and art (especially Romantic art) are two fundamentally different enterprises. Others believe that in Holmes these two enterprises are at some level fused or combined, yet their comments also assume that what makes Holmes so striking and unusual is precisely this blending of what we normally find opposed. In the words of John Cawelti, Holmes "is the stereotype of the rational, scientific investigator, the supreme man of reason. Yet, at the same time, his character paradoxically incorporates basic qualities from a contrary stereotype, that of the dreamy romantic poet, for Holmes is also a man of intuition, a dreamer, and a drug-taker, who spends hours fiddling aimlessly on his violin." Like Dupin before him, Holmes synthesizes "the poet's intuitive insight with the scientist's power of inductive reasoning."[6]

It is not difficult to find support for such a position. In a famous passage, Watson is surprised by Holmes's ability, while in the middle of a case, to give way to a state of languorous enjoyment at a violin recital: "All the afternoon he sat in the stalls wrapped in the most perfect happiness, gently waving his long, thin fingers in time to the music, while his gently smiling face and his languid, dreamy eyes were as unlike those of Holmes, the sleuth-hound, Holmes the relentless, keen-witted, ready-handed criminal agent, as it was possible to conceive. In his singular character the dual nature alternately asserted itself, and his extreme exactness and astuteness presented, as I have often thought, the reaction against the poetic and contemplative mood which occasionally predominated in him."[7] Here we have the "alternate assertion" of a "dual nature," active versus contemplative, the exact versus the poetic. Yet Watson is often a notoriously bad observer and reasoner, and here, as he continues, this neat dichotomy is quickly undermined, for when Holmes is engaged on an interesting case, his "brilliant reasoning power would rise to the level of intuition." The exact no longer "reacts against" the poetic, but joins it.

## SHERLOCK HOLMES AS SCIENTIST

In his biography of Doyle, Pierre Nordon says that "each period makes its own image of the scientist."[8] Before we meet Holmes—indeed, before Watson meets Holmes—he is described to us not as a detective but as a desultory sort of science student. Young Stamford tells Watson that Holmes is "working at the chemical laboratory up at the hospital" but "has never taken out any systematic medical classes"; an "enthusiast in some branches of science," he is "well up" in anatomy and "a first-class chemist," yet his "eccentric" studies have principally enabled him to "amass a lot of out-of-the-way knowledge which would astonish his professors" (STUD, 1, 1). When Stamford introduces the two men, Holmes has just completed a successful chemical experiment, and from time to time throughout the saga Watson will refer to Holmes's working all night on a chemical analysis. But our exposure to Holmes's scientific methods comes not through his science but through his detection. We meet him at the moment he has *completed* a successful experiment, just as we watch him apply the results of his research on the various types of cigar ash but are not told of the process of that research. In fact, when Watson is first treated to a display of Holmes's detective abilities in "A Study in Scarlet"—a display that in part depends on the identification of Trichinopoly cigar ash—Watson's awestruck compliment that Holmes has "brought detection as near an exact science as it ever will be brought in the world" (1, 4) is recapitulated almost exactly in Doyle's recollection of his determination that his detective would "reduce this fascinating but unorganized business to something nearer to an exact science."

Since Doyle's conception of Holmes is that of the scientific detective, the image of Holmes provides us with a popular image of the scientist and his methods. Nordon goes so far as to claim that Holmes stimulated public interest in science, and Ian Ousby contends that "the underlying intention of the stories" was "to popularize scientific method."[9] While we need not go so far as this, we do need to ask just what image Doyle gives us of the scientist, and just what is meant by the "scientific method" that Holmes practices.

In his early and influential study *Le « Detective Novel » et l'influence de la pensée scientifique* (1929), Régis Messac considered the question of Holmes's methods and concluded that they could be considered scientific only in a very vague and general sense. Holmes's "deductions" are not really deductions at all, but neither are they inductions. Of Holmes's comments about analytic reasoning, Messac says that this is not science but pseudoscience. Although Holmes is a man armed with logic and reason, this gives to the stories no specifically identifiable scientific method.[10] Subsequent critics have tended to follow Messac's line, arguing that Holmes is part of the general scientific spirit of the age, and that his methods represent confidence in a scientific method assumed to be monolithic.[11]

In recent years, however, there has been a significant resurgence of interest in Holmes's methods, a resurgence dominated by semioticians. Led by Umberto Eco and Thomas Sebeok, these critics have argued that Holmes's methods correspond to C. S. Peirce's concept of abduction.[12] In contrast to these two widely different approaches, I want to contend that we can be both more and less specific about Holmes's methods. More, because we have a better sense of what the late Victorians would have meant by a "scientific spirit," or, more appropriately, of the tensions that such a term contained. Less, because the extraordinary debate over scientific method makes the identification with a single view highly problematic. Rather, we can see the contradictions contained within Holmes's methods (whether spoken or actual), like the contradictions within his "dual nature," as tensions which were absorbed from and participated in the cultural debate about scientific method.

## SHERLOCK HOLMES THE BACONIAN

I want to begin my analysis of Holmes's method by identifying and separating various strands of Holmes's methodology. As I proceed, however, I will try to give a sense of how each strand, rather than standing alone, interacts dynamically with—here supporting, there contradicting, there modifying—other actions or methodological claims by the great detective.

Central to Holmes's methods are a number of features that hearken back to the naive Baconianism against which the century reacted, features that in the second half of the century were also associated with extreme versions of positivist thought. Chief among them was the emphasis on facts. If "when you have eliminated the impossible, whatever remains, however improbable, must be the truth" is the most famous of Holmes's methodological precepts, the second is that "[i]t is a capital mistake to theorize before you have all the evidence. It biases the judgment" (STUD, 1, 3). Both injunctions are voiced in the earliest stories and repeated in later ones. Holmes, quite naturally, considers himself to be one of the few people who can resist this unconscious temptation, for he tells Watson that "I make a point of never having any prejudices, and of following docilely wherever fact may lead me" (REIG). Holmes's notion, like that of the naive Baconian, is that fact-collecting is not itself a theory-laden activity, and therefore the fact-collector can achieve pure objectivity by virtually effacing himself from the process. In "The Adventure of the Cardboard Box," Holmes extends the concept of a Lockean tabula rasa to claim that a detective or a scientist can wipe his mental slate clean with each new investigation: "We approached the case . . . with an absolutely blank mind, which is always an advantage. We had formed no theories. We were simply there to observe. . . ." The vision of the active and energetic Holmes "following docilely," "simply" observing, is a jarring one, yet such a view is crucial to the notion that the truth at which the detective arrives has nothing to do with luck on the one hand, personal genius on the other. When personality infuses the process, theories become a personal creation to which the detective is apt to become too much attached, ignoring the facts that lead inexorably to the true theory.

The empirical ability of the blank mind to collect facts and then form theories separates the activities of observation and theory formation into distinct and consecutive pieces of the detective process. In its extreme form, this separation can lead to a division of labor in which observation and theory can be carried out by different people. It is precisely this division of labor that Holmes employs in "The Hound of the Baskervilles," sending Watson to Devonshire to collect facts while he remains in London to theorize: "'I will not bias your mind by suggesting theories or suspicions, Watson, . . . I wish you simply to report facts in the fullest possible manner to me, and you can leave me to do the theorizing'" (6).

It is crucial to see that "theorizing," whether in the naive Baconian or Holmesian scheme, places a premium on thoroughness and common sense rather than intuitive insight and genius. The scientist must be meticulous, whether in formulating the initial list of possible explanations for a phenomenon, or in eliminating items from that list, or in moving from one inductively established truth to the next. Thus, as Robert Leslie Ellis argued in his preface to Bacon's philosophical works, the Baconian system of elimination is designed not only to generate certainty, but to be applied by anyone, virtually without regard to genius. The patience and diligence of gradual elimination are also crucial to Holmes's attack on the mysteries presented in "A Study in Scarlet." At the end of the case he treats Watson to an extensive, step-by-step explanation of both his investigation and his reasoning. "'By the method of exclusion,'" he tells Watson, "'I had arrived at [the] result, for no other hypothesis would meet the facts'" (2, 6). In "The Adventure of the Blanched Soldier," Holmes determines simply from the statement of the case that there is "a very limited choice of alternatives"; from this point he begins asking the questions that will "narrow down the possible solutions." Indeed, that oft-repeated cardinal rule—"when you have eliminated the impossible, whatever remains, however improbable, must be the truth"—obviously owes much to the Baconian tradition in scientific method. There is only one true solution, so the gradual elimination of all but one possibility from a carefully prepared list yields absolute certainty.

## SHERLOCK HOLMES AND THE HYPOTHETICO-DEDUCTIVE METHOD

When Holmes sends Watson to Devonshire in "The Hound of the Baskervilles" with the injunction that he "simply report facts," the ever-practical Watson asks, "'What sort of facts?'" (6). This question goes to the heart of what the century had seen as one of the chief problems with the naive Baconian system: how can someone "collect facts" unless he already has a theory in mind that suggests what to look for, where to look for it, and how to interpret what is found? Isn't theory, hypothesis, speculation—and thus the individual imagination—an indispensable feature of scientific method?

Holmes's answer, echoing that of nineteenth-century writers on method, is yes. In fact, in Holmes's eyes, it is the official police who are the narrow-minded and unsuccessful Baconians in contrast to his more open, boldly theoretical methods. He commends the inspectors of Scotland Yard for their meticulous collection of facts but in the same breath criticizes their imaginative limitations: they "'lead the world for thoroughness and method'" but suffer "'an occasional want of imaginative intuition'" (3GAR); though "'excellent at amassing facts, they do not always use them to advantage'" (NAVA). This leads the men of the Yard into two types of dilemmas. On the one hand, they are often unable to make any sense of a case at all, for their objective precepts provide them with no starting point, no way to wade through the mass of facts and select those which are truly important. On the other hand, they take their evidence at face value, assuming that the "facts" are self-evidently true and therefore require no interpretation; they trust uncritically the testimony of the senses.[13]

Not surprisingly, the official police are equally critical of Holmes's methods. Naive Baconians that they are, they have little use for the displays of an unrestrained imagination. Inspector Gregson tells Holmes that "'[w]e want something more than mere theory and preaching'" (STUD, 1, 7). Athelny Jones confides to Watson that Holmes is "'irregular in his methods and a little quick perhaps in jumping at theories'" (SIGN, 9); he condescendingly remarks that although Holmes "'has the makings of a detective in him,'" his methods are "'just a little too theoretical for me'" (REDH). Inspector Lestrade grows sarcastic when Holmes questions his solution of the case: "'We have got to the deductions and the inferences,' said Lestrade, winking at me. 'I find it hard enough to tackle facts, Holmes, without flying away after theories and fancies'" (BOSC).

Yet the necessity of theory is made evident even before we see Holmes on a case. After he moves in with his new roommate, Watson grows increasingly curious about the nature of Holmes's mysterious occupation. When Holmes tells Watson that he stocks his "brain-attic" only with the "furniture" which will "help him in doing his work," Watson tries to determine from a list of Holmes's eclectic accomplishments what that "work" must be: "I pondered over our short conversation . . . and endeavoured to draw my deductions from it. . . . I enumerated in my own mind all the various points upon which he had shown me that he was exceptionally well informed. I even took a pencil and jotted them down. . . . When I had got so far in my list I threw it into the fire in despair. 'If I can only find what the fellow is driving at by reconciling all these accomplishments, and discovering a calling which needs them all,' I said to myself, 'I may as well give up the attempt at once' " (STUD, 1, 2). Despite his "enumeration" of a list from which he can generate another list of possible professions, the systematic elimination of which will in its turn yield the solution, Watson's "deductions" fail miserably. In part this is because Holmes is unique—" 'I have a trade of my own. I suppose I am the only one in the world. I'm a consulting detective, if you can understand what that is' "—but more important it is due to an inadequacy of method: a collection of facts doesn't lead simply to a solution without the mediation of inference, hypothesis, intuition. Even in "The Hound of the Baskervilles," Holmes only pretends to be back in London spinning theories from the facts sent to him by Watson—he is in fact himself in Devonshire, busy at both collecting facts and spinning theories.

Yet just what role the imagination plays in Holmes's methods is not necessarily easy to grasp. A succinct example of the difficulties encountered appears in Holmes's summary of his investigations in "The Adventure of the Cardboard Box." As already noted, Holmes claims that he and Watson approached the case with a blank mind, with no preconceived theories, simply ready to observe "and to draw inferences from our observations." This drawing of inferences begins to stress the need for interpreting facts, that facts do not simply reveal more general truths of their own accord, but it is still something that occurs after the collection of facts. Holmes goes on, however, to cast his account in visual terms, first asking rhetorically what they observed, what they physically saw, and then what they inferred, what they mentally saw. " 'What did we see first? A very placid and respectable lady, who seemed quite innocent of any secret, and a portrait which showed me that she had two younger sisters.' " This is simple, physical observation, but the inference from it takes an imaginative leap: " 'It instantly flashed across my mind that the box might have been meant for one of these.' " The "instant flashing" is more intuitive than logical, and clearly eschews the gradual process demanded by Bacon. Yet Holmes immediately limits his speculation as requiring confirmation: " 'I set the idea aside as one which could be disproved or confirmed at our leisure.' "

This is the pattern for the entire explanation—a step-by-step account, suffused with the language of observation and logical inference, but studded also with that of sudden insight and revelation ("'at once a whiff of the sea was perceptible,'" "'I at once saw the enormous importance of the observation'").

What Holmes demands first is a "framework of fact" on which the imaginative elements of his method—the drawing of inferences, the creation of hypotheses—can build: "'It is one of those cases where the art of the reasoner should be used rather for the sifting of details than for the acquiring of fresh evidence. . . . [W]e are suffering from a plethora of surmise, conjecture, and hypothesis. The difficulty is to detach the framework of fact—of absolute, undeniable fact—from the embellishments of theorists and reporters. Then, having established ourselves upon this sound basis, it is our duty to see what inferences may be drawn and what are the special points upon which the whole mystery turns'" (SILV). Holmes here still separates the collection of facts from the development of theory, but he does acknowledge that the selection of the "special points," the crucial facts, is a part of his methodology requiring both logic and imagination. Elsewhere, however, he admits that even observation, and hence the collection of "absolute, undeniable fact," must be directed by theory. On one of many occasions when he chides Watson for seeing but not observing, Holmes tells his friend that "'[y]ou did not know where to look, and so you missed all that was important'" (IDEN). On another occasion, Holmes finds a match in the mud, much to the dismay of Inspector Gregory, who expresses his annoyance at having "overlooked" it; "'It was invisible,'" Holmes (partly) reassures him. "'I only saw it because I was looking for it'" (SILV). Thus there seem to be two classes of facts. There are the obvious, incontrovertible facts that form the basis of the situation—a racehorse is missing, his trainer is found bludgeoned to death on the moor, a stranger appeared at the stables the night before, a stable boy was drugged. This is Holmes's "framework of fact." But other facts, vital to resolving the meaning of this factual framework, can be found only through observation that is directed by the construction of a theory. One finds matches in the mud because one expects to find matches in the mud.

In spite of his advice never to theorize in advance of data, Holmes knows that it usually becomes necessary to construct a preliminary hypothesis, and that this hypothesizing is valid so long as it covers all the facts and is subjected to confirmation: "'One forms provisional theories and waits for time or fuller knowledge to explode them. A bad habit, . . . but human nature is weak'" (SUSS). When hired to absolve someone accused of a crime, Holmes is of course willing to start from the assumption that the person's story is true, whatever the evidence may be. But on other occasions hypotheses are necessary simply to get the investigation started, without regard to truth. In "The Valley of Fear" Holmes claims that "'there should be no combination of events for which the wit of man cannot conceive an explanation. Simply as a mental exercise, without any assertion that it is true, let me indicate a possible line of thought. It is, I admit, mere imagination; but how often is imagination the mother of truth?'"

"Mere" imagination is balanced by its ability to engender truth, yet elsewhere Holmes does not even qualify the value of the imagination with "mere." When Watson can determine nothing about a man from the observation of his hat, Holmes chides him for being "'too timid in drawing your inferences'" (BLUE). When Holmes develops a theory, carries out his investigation in accordance with it, and finds evidence that confirms it, he invites Watson to "'[s]ee the value of the imagination. . . . We imagined what might have happened, acted upon the supposition, and find ourselves justified'" (SILV). Such imagination is especially important because there are often multiple possible explanations to fit a group of facts, and it is essential that all be assembled before the process of elimination begins. In a significant qualification to the maxim that when you have eliminated the impossible, whatever remains must be the truth, Holmes acknowledges that "'[i]t may well be that several explanations remain, in which case one tries test after test until one or other of them has a convincing amount of support'" (BLAN). In this formulation, the process is mechanical and impersonal (one tries test after test) but is constructive (one doesn't simply eliminate bad explanations, but builds support for the best explanation). On another occasion, the essentially negative Baconian process of constructing a list and then eliminating items from it is presented, but this process of elimination is seen not as mechanical but as creative, with Holmes "devising" his explanations: "'I have devised seven separate explanations, each of which would cover the facts as far as we know them. But which of these is correct can only be determined by the fresh information which we shall no doubt find waiting for us'" (COPP).

Yet in spite of his caution, there are times when Holmes is too bold in drawing inferences, when he theorizes without sufficient data, when he even relies on intuition rather than factual grounding. In "The Adventure of the Speckled Band" he admits to Watson that he had "'come to an entirely erroneous conclusion'" because he had immediately associated the word "band" with gypsies rather than a snake. At their most spectacular, Holmes's methods permit him to solve a case without making any observations or interviewing the principals—observation and interrogation merely confirm his conclusions. On other occasions, however, the necessity of employing hypotheses can set Holmes off on a false scent. Yet these displays of Holmes's fallibility are balanced by those cases, largely ignored by students of Holmes's methods, where he depends on intuition. In one such case, after he regales Watson with an account of his unsuccessful investigations, Holmes is nonetheless certain that, in spite of the evidence, he will save his client: "'So, my dear Watson, there's my report of a failure. And yet—and yet—' he clenched his thin hands in a paroxysm of conviction—'I *know* it's all wrong. I feel it in my bones'" (NORW). Similarly, in "The Adventure of the Abbey Grange," Holmes is certain that the apparently resolved case has not been resolved at all; pulling Watson from the train at one of the stops on their return journey to London, he tells his companion that "'I am sorry to make you the victim of what may seem a mere whim, but on my life, Watson, I simply can't leave that case in this condition. Every instinct that I possess cries out against it. It's wrong—it's all wrong—I'll swear it's wrong. And yet the lady's story was complete, the maid's corroboration was sufficient, the detail was fairly exact. What have I to put up against that?'"

This reliance on instinct and intuition in the face of facts is troubling, given that both he and Watson elsewhere enforce the traditional view of instinctual knowledge as a feminine trait, a view in which that knowledge is both praised as genuine and criticized as subjective, emotional, inferior. In "The Boscombe Valley Mystery," for example, Watson notes that Miss Turner, "with a woman's quick intuition," is able to tell, without an introduction, which of the two of them is Holmes. She then pours out her heart, telling the detective that she "knows" her friend James McCarthy could not be the murderer of his father. When Holmes expresses his view that it is "very probable" that McCarthy is innocent, Inspector Lestrade, who is also present and is of course sure that the case against McCarthy is solid, attempts to dampen Miss Turner's hopes by noting that "'I am afraid my colleague has been a little quick in forming his conclusions.'" Although Miss Turner turns out to be right—McCarthy is not the murderer—her "knowledge" is immaterial to the case. From reading the evidence, Holmes thinks it "probable" that McCarthy is innocent before he even meets Miss Turner; what he gains from her is not any sense of conviction about McCarthy's character, but information. Holmes's intuition, though linked to Miss Turner's by Lestrade's comment, is distinguished from hers and made authoritative by its "logical," "objective" basis and its ability to generate physical evidence. But more crucially, Holmes's methods enable him to see what Miss Turner's intuition cannot, that her own father is the murderer.

Feminine intuition, then, is accorded very limited status in comparison to the implicitly male methodology of science. Although he tells one of his female clients that "'I have seen too much not to know that the impression of a woman may be more valuable than the conclusion of an analytical reasoner'" (TWIS), Holmes more often asserts that "the impression of a woman" impedes the analytical process. Whereas his intuition is grounded in fact or at least has a professional basis (his experience with other cases), and is subject to confirmation, that of women is of only marginal utility. In this case, the cool and relatively passive male reason is active in comparison with its female counterpart; Miss Turner's intuitive knowledge, though highly personal and aggressively asserted, requires Holmes's impersonal and measured activity if practical results are to be achieved.

Intuition and emotion, though at some level essential to Holmes's method, must be vigorously circumscribed and controlled. Indeed, on that occasion when Holmes's "every instinct" cries out that the case is wrong, what has made it go wrong, he tells Watson, is that "'the lady's charming personality'" has "'warped our judgment'": "'if I had not taken things for granted, if I had examined everything with the care which I should have shown had we approached the case *de novo* and had no cut-and-dried story to warp my mind, should I not then have found something more definite to go upon? Of course I should'" (ABBE). Giving in to imagination—like giving in to woman herself—is a fatal temptation. In explaining to Watson that he would never marry "'lest I bias my judgment,'" Holmes says that "'love is an emotional thing, and whatever is emotional is opposed to that true cold reason which I place above all things'" (SIGN, 12).

According to Watson, Holmes "never spoke of the softer passions, save with a gibe and a sneer" (SCAN). His admiration for Irene Adler, the woman who bests him in "A Scandal in Bohemia," is based on her ability to outwit him; "it was not," Watson is at pains to tell us, "that he felt any emotion akin to love" for her. All emotions, but especially love, "were abhorrent to his cold, precise but admirably balanced mind," for emotions have the potential to disrupt that balance, to upset what is here perceived as a purely mechanical reasoning process: "for the trained observer to admit such intrusions into his own delicate and finely adjusted temperament was to introduce a distracting factor which might throw a doubt upon all his mental results. Grit in a sensitive instrument, or a crack in one of his own high-power lenses, would not be more disturbing than a strong emotion in a nature such as his." Though Irene Adler is *the* woman to him, Holmes must circumscribe his admiration within impersonal, objective limits if he is to safeguard his methods from corruption. Like the imagination, she can be admired—but not too much. A powerfully sexual being, with "the face of the most beautiful of women," one-time lover of the crown prince of Bohemia, she is still, according to the now king, " 'an adventuress,' " and according to Watson, "a woman of dubious and questionable memory."

It is at such times as these that science and scientific method are most closely associated with the inhuman and impersonal. If to be a scientist means to be so utterly unemotional, so cut off from the world of sexual love, then surely science is incomplete. If a man is "a calculating machine," then he is not really a man. When describing Holmes to Watson, Stamford judges Holmes's "passion for definite and exact knowledge" to be "a little too scientific for my tastes" and says it "approaches to cold-bloodedness" (STUD, 1, 1). Holmes's failure to observe the attractions of Mary Morstan (the woman Watson will marry) in "The Sign of Four" elicits from Watson the cry (also echoed by Doyle in *Memories and Adventures*) that he is " 'an automaton—a calculating machine,' " that there is " 'something positively inhuman in you at times' " (2). Unperturbed, Holmes responds that " '[i]t is of the first importance not to allow your judgment to be biased by personal qualities. A client is to me a mere unit, a factor in a problem. The emotional qualities are antagonistic to clear reasoning.' "

Such comments tend to be packed into the earlier stories.[14] In one of the last of them, Watson displaces this view of Holmes as a "calculating machine" from himself or from Holmes onto other people. From his reference to Holmes's composed face as that "which had made so many regard him as a machine rather than a man" (CROO), it is clear that Watson now recognizes that Holmes is not simply a machine, and that he believes such a view amounts to a caricature of both Holmes's character and his methods. No one who knows him well, Watson is claiming, could possibly consider him inhuman. Suggestions of this appear quite early, for even in the first short story, Holmes's mind, though "cold and precise," is also "finely balanced."

## HOLMES AND NINETEENTH-CENTURY SCIENCE: CHEMISTRY, CUVIER, HUXLEY, AND TYNDALL

If Joe Bell's diagnostic skills provided part of the model for Holmes's methodology, it is in some ways a relatively restricted part: Conan Doyle's anecdotes about Bell most obviously apply to those showy displays at the beginning of almost every story, at the expense of Watson or the client, when Holmes mysteriously "knows" all about someone he has never met.[15] Holmes uses these displays to impress his clients with his abilities, yet he is quick to point out to Watson that these flashes of intuition are actually the result of a gradual reasoning process, a process that practice has enabled him to hone and compress. Thus "'it is not really difficult to construct a series of inferences, each dependent upon its predecessor and each simple in itself. If, after doing so, one simply knocks out all the central inferences and presents one's audience with the starting-point and the conclusion, one may produce a startling, though possibly a meretricious, effect'" (DANC). Such is exactly what happens in their first meeting, when Watson is amazed by Holmes's "knowledge" that he has been in Afghanistan. As Holmes explains, "'[f]rom long habit the train of thoughts ran so swiftly through my mind that I arrived at the conclusion without being conscious of intermediate steps. There were such steps, however'" (STUD, 1, 2). Watson soon comes to echo Holmes's claims: "I had no keener pleasure than in . . . admiring the rapid deductions, as swift as intuitions, and yet always founded on a logical basis . . ." (SPEC).

Yet if Holmes is willing to provide the logical steps of these little displays at the beginning of the story, or even of the whole case at the end, he is careful not to explain—at least not fully—while a case is under way. After a partial explanation of the mysteries in "A Study in Scarlet," Holmes demurs when Watson asks for more: "'I'm not going to tell you much more of the case, Doctor. You know a conjurer gets no credit when once he has explained his trick; and if I show you too much of my method of working, you will come to the conclusion that I am a very ordinary individual after all'" (1, 4). Like both Doyle and Watson in their roles as authors, Holmes must both reveal and conceal. Like Joe Bell, he enjoys the dramatic effect, but it is crucial if his results are to be accepted that his methods not be too mysterious, the product of chance or idiosyncracy or intuitive genius—in short, Holmes's methods and results, like those of the scientist, must be reproducible.

Likening his procedure to a "conjurer's trick" clearly raises potentially uncomfortable associations. Better models, similar to that of Joe

Bell but with more far-reaching applications, dramatic and yet with a fairly high degree of authority, can be found in the methodologies of natural sciences like chemistry and paleontology/geology/comparative anatomy. As we have already seen, we first meet Holmes just as he has completed a chemical experiment. Immediately thereafter, we hear described what Holmes calls "the Science of Deduction and Analysis." "Analysis"—which obviously has both logical and chemical connotations—is associated by Holmes with the deductive process of reasoning from effect to cause:

"the grand thing is to be able to reason backward. That is a very useful accomplishment, and a very easy one, but people do not practise it much. In the everyday affairs of life it is more useful to reason forward, and so the other comes to be neglected. There are fifty who can reason synthetically for one who can reason analytically. . . . Most people, if you describe a train of events to them, will tell you what the result would be. They can put those events together in their minds, and argue from them that something will come to pass. There are few people, however, who, if you told them a result, would be able to evolve from their own inner consciousness what the steps were which led up to that result. This power is what I mean when I talk of reasoning backward, or analytically." (STUD, 2, 7)

Given an unknown compound (the crime), Holmes analyzes it, takes it apart in order to see what it is made of. Yet the chemical analogy is somewhat faulty when applied to a crime, because analysis has little to say about the history of the crime, the mechanism of its production. Thus, though logical and systematic, analysis also involves imaginative creativity, the ability of the detective/chemist to "evolve" the mechanism from his "own inner consciousness."

A similar argument is made in connecting Holmes's methods to those of the paleontologist and comparative anatomist. "'The ideal reasoner,'" says Holmes, "'would, when he had once been shown a single fact in all its bearings, deduce from it not only all the chain of events which led up to it but also all the results which would follow from it. As Cuvier could correctly describe a whole animal by the contemplation of a single bone, so the observer who has thoroughly understood one link in a series of incidents should be able to accurately state all the other ones, both before and after'" (FIVE). Holmes here attaches a scientist's name to a process already described in his article "The Book of Life," from which Watson reads in "A Study in Scarlet": "From a drop of water, . . . a logician could infer the possibility of an Atlantic or a Niagara, without having seen or heard of one or the other" (1, 2). And this sense of the interconnectedness of natural phenomena is of course crucial to his claim, frequently repeated in the stories, that his method "is founded upon the observation of trifles" (BOSC). From these trifles, Cuvier's method constructs both a historical record and a mechanism, aspects less apparent in the chemical analogy. The animals Cuvier dealt with were extinct, the bones he dealt with were fossilized fragments. Yet out of these fragments, Cuvier reconstructed the appearance of an entire animal. Such reconstructions depended heavily on the use of analogical reasoning, on the knowledge that, for example, a certain length of foot bone corresponded to a certain length of leg, which in turn was related to overall size, gait, posture, etc. Just as clearly, to make such analogies required extensive familiarity with a wide variety of creatures, both living and extinct, from which to draw inferences.

The methodology of Cuvier thus corresponds with what is Holmes's most thorough exposition of the methods of the detective. In "The Sign of Four," Holmes speaks of a *French* disciple who "'has come rather to the front lately in the French detective service'" and is "'translating my small works into French'" (1). This disciple has "'all the Celtic power of quick intuition'" but is "'deficient in the wide range of exact knowledge which is essential to the higher developments of his art.'" Indeed, he "'possesses two out of the three qualities necessary for the ideal detective. He has the power of observation and that of deduction. He is only wanting in knowledge, and that may come in time.'" It is this three-pronged description that most fully articulates Holmes's method by combining the need for observation of fact and encyclopedic knowledge with an intuitive but logical brilliance. Like Cuvier, Holmes—often with very little to start from—uses his powers of observation and his knowledge of the structure of other cases as a foundation on which to build his inferences, moving, as he so often says, from effect to cause, until he can "construct the whole animal from a single bone." And like Cuvier's reconstructions, Holmes's are historical, using present evidence to determine exactly how a past event—the crime—must have occurred.

The models of the chemist and the paleontologist are also useful for reconciling what otherwise appears as a contradiction. Holmes often claims that a case's outré features are what make it solvable, yet he also argues that there are no truly unique crimes, and hence that an encyclopedic historical knowledge of crime permits the detective to solve cases quickly and easily. At one level, the outré and the everyday are clearly related, for only with such extensive historical knowledge can a detective know what is truly unique or unusual about a case and what is not. But both situations are also linked by the difficulty in knowing where and how to start the investigation. In "A Study in Scarlet," Holmes specifically links the outré with analytic reasoning: the outré is the place from which the detective begins his deductions. When a case lacks outré features, however, the selection of a starting point is much more problematic. It is here that Holmes relies, when possible, on his knowledge of previous cases. Since crimes do not exactly recapitulate themselves, one crime cannot be used to supply an immediate solution to another, but the earlier crime does suggest analogies that can initiate the process of theory formation. Holmes's insistence that outré cases are easier to solve is thus not only consistent but accurate, for in the outré case, as in the chemical analysis, there is no question of what is to be taken apart and how this process is to be begun. The paleontological model is more appropriate because most cases do not lend themselves to such a mechanical approach and because the selection and development of the important facts are more difficult. Like the paleontologist faced with a tantalizing fragment of bone that seems not to fit any existing scheme, Holmes uses what he does know as a source of analogies by which he can better define both what is similar and what is different, with the result that the new fragment becomes incorporated into—though altering as it does so—the previously existing scheme.

Another figure in the method of Cuvierian reconstruction is T. H. Huxley, who serves as a pivot point in a matrix of methodological connections. Recalling his rejection of Catholicism and of Christianity during his student days, Doyle remarked that "[i]t is to be remembered that these were the years when Huxley, Tyndall, Darwin, Herbert Spencer and John Stuart Mill were our chief philosophers, and that even the man in the street felt the strong sweeping current of their thought, while to the young student . . . it was overwhelming."[16] For Huxley, Cuvier's method—and, indeed, the method of all the historical sciences—was "the method of Zadig."[17] Zadig's story is an ancient one, but it came to Huxley through Voltaire (1747). Set in Babylon, the story concerns the disappearance of the king's best horse and the queen's pet spaniel. In the process of searching for the missing animals, the king's grand huntsman and the queen's eunuch come across Zadig, a sort of retired wise man now living in seclusion in order to study nature. Zadig hasn't seen the animals and doesn't even know they exist until he's told they're missing, but when the servants question him, he is able to describe the animals with remarkable accuracy. When Zadig insists that he's never seen the animals and doesn't know where they are, the servants assume that he is responsible for the disappearance. He is arrested and taken to court, where he explains how he inferred what the animals looked like from the signs he had seen of their passing (the distance between the horse's hoofprints indicated its length, broken tree branches its height, etc.).

In Huxley's commentary, Zadig "sharpened his naturally good powers of observation . . . [and] at length acquired a sagacity which enabled him to perceive endless minute differences among objects which, to the untutored eye, appeared exactly alike." The "foundation of all Zadig's arguments," says Huxley, is a "commonplace assumption upon which every act of our daily lives is based, that we may conclude from an effect to a pre-existence of a cause competent to produce that effect." The similarity to Holmes's methods is obvious, but it becomes even more so as Huxley uses Zadig's method for the model of the methodology of the historical sciences: "the rigorous application of Zadig's logic to the results of accurate and long-continued observation has founded all those sciences which have been termed historical . . . because they . . . strive towards the reconstruction in human imagination of events which have vanished and ceased to be" (*Collected Essays*, 4:3, 7, 9).

The bulk of Huxley's essay is devoted to discussion of examples from a variety of sciences, but especially to Cuvier and comparative anatomy/paleontology. After noting Cuvier's ability to reconstruct an animal from a single fossil, he tells a story of Cuvier examining part of an exposed skeleton in a quarry and determining from the teeth and jaw what type of an animal it must have been and what the as yet unexposed pelvis would look like. Cuvier is for Huxley the modern scientific exemplar of Zadig's method, a fact reinforced in the quote from Cuvier that stands as the essay's epigraph: "Une marque plus sûre que toutes celles de Zadig."

Cuvier's presence is also very much in evidence in an earlier discussion of scientific method, "On the Educational Value of the Natural History Sciences" (1854). In this essay, Huxley argues that "the methods of all science whatsoever" are (1) observation of facts, including artificial observation, or experiment; (2) comparison and classification of facts into general propositions; (3) deduction; (4) verification (*Collected Essays*, 3:52–53). It is easy to see how closely this corresponds to the general activities of Holmes, where a provisional theory is constructed on the basis of a framework of fact and of comparison with previous cases, and from this theory deductions are made and then tested. To make the matrix denser, Huxley acknowledges his own debt to another of the "chief philosophers" whom the young Doyle found "overwhelming": "I need hardly point out my obligations to Mr. J. S. Mill's *System of Logic*, in this view of scientific method" (*Collected Essays*, 3:56n). But even more extraordinary is the following passage, in which Huxley associates Cuvier's method with that of the detective: "Science is, I believe, nothing but *trained and organized common sense*. . . . [T]he vast results of Science are won by no mystical faculties, by no mental processes, other than those which are practised by every one of us, in the humblest and meanest affairs of life. A detective policeman discovers a burglar from the marks made by his shoe, by a mental process identical with that which Cuvier restored the extinct animals of Montmartre from fragments of their bones."[18]

Yet there is also a difference between the historicity of Holmes's and Cuvier's reconstructions. The paleontologist and comparative anatomist take the evidence of the present (a fossil) and construct from it an animal that lived at some point in the historical past. The emphasis is essentially static, centered on two moments in time. Cuvier was not interested in the development of the animals he reconstructed, how they came into being and later came to be extinct; indeed, the pre-Darwinian assumption was that such species sprung into being fully developed, filling a niche in their environment until that environment changed and they, unable to change with it, became extinct. Hence the fascination with Cuvier's method is much more spatial than temporal—what interests Holmes is what interested Cuvier, the ability to construct an entire animal out of a single bone. But Holmes's "ideal reasoner" is concerned with chronological, not spatial, reconstructions. From Holmes's "single fact," corresponding to Cuvier's "single bone," the ideal reasoner constructs not a series of spatial relationships but "a series of incidents," a "chain of events." And crucially, this chain is not limited to the events that have led up to this single fact, but also extends to "all the results which would follow from it." The method both explains the past and predicts the future.

This ability to deduce from a single fact and general laws the entire temporal course of a complete spatial system is more closely associated with mechanics and physical astronomy. With a complete description of the universe at one moment in time and the complete physical laws governing the motion of its members, Laplace had claimed earlier in the century that a scientist could, ideally, describe the universe's past, present, and future. Such views, of course, are highly mechanistic, deterministic, and materialistic. As a model for Holmes's methods, they fail to capture the intuitive element of his thought; worse, they were repulsive to Doyle, who, despite his interest in science, was as critical of the extreme form of these tendencies as his Romantic forebears: "How a man of science can be a materialist is . . . amazing to me."[19] By yoking this Laplacean legacy with Cuvier's, however, Doyle fuses the mechanistic materialism of the physical sciences with the more implicitly creative activities of the historical sciences. It is yet another example of trying to find an adequate means of portraying Holmes's methods as being both mechanical and creative, factual and speculative, objective and subjective, Baconian and hypothetical-deductive. But if the method of Zadig, Cuvier, and Huxley also does not quite capture the totality of Holmes's enterprise, and if the combination of Cuvier and Laplace raises some disquieting implications from the Laplacean side, that does not mean that the physical sciences could not offer a method, based on but modifying that of Laplace, that did capture that totality. The key, I think, is another of Doyle's "chief philosophers": John Tyndall.

In "The Hound of the Baskervilles," Holmes analyzes an anonymous warning note sent to Sir Henry Baskerville. He deduces that the note's words were cut out of a particular *Times* article with a pair of nail scissors by an educated man trying to conceal his handwriting and hence someone known, or likely to become known, to Sir Henry. The somewhat haphazard pasting of the words onto the notepaper suggests to Holmes that the man may have feared interruption, but to this Dr. Mortimer says that "[w]e are coming now rather into the region of guesswork" (4). Holmes's reply is telling: "Say, rather, into the region where we balance probabilities and choose the most likely. It is the scientific use of the imagination, but we have always some material basis on which to start our speculation." Whether or not Doyle employed Tyndall's phrase consciously, his use of it cannot be accidental, especially given its points of contact with Holmes's conception of method.

To many of his contemporaries, Tyndall was the arch materialist. In an 1868 address to the British Association, later reprinted as "Scientific Limit of the Imagination" and "Scientific Materialism," he does Paley one better by comparing not just nature, but the human body and human mind, to a watch. The task of physical science is to uncover the "inner mechanism" of nature and the "store of force . . . [that] set[s] that mechanism going." Of the body, Tyndall says that many of its parts are "obviously mechanical," and therefore that "given the relation of any molecule of the body to its environment, its position in the body might be determined mathematically." "With the necessary molecular data," says Tyndall, "the chick might be deduced as rigorously and as logically from the egg as the existence of Neptune was deduced from the disturbances of Uranus. . . ." Tyndall moves from egg to chick, from molecule to body, as easily as Holmes moves from a drop of water to the Atlantic, or Cuvier from fossil fragment to complete skeleton. Furthermore, even mental states can be explained by molecular processes: "thought . . . has its correlative in the physics of the brain."[20]

Yet Tyndall insists that his form of scientific materialism depends on rather than excludes the mysteries of the imagination. He contends his "main object" is "to dissipate the repugnance, and indeed terror, which in many minds are associated with the thought that science has abolished the mystery of man's relation to the universe." To say that "thought has its correlative in the physics of the brain" is not to explain the mystery of why this correlation exists (1:405–7). Similarly, the method of the physical scientist cannot be described in simply mechanical terms. Although "accurate observation and experiment" are essential to science, they are only the beginning of the process; rejecting naive Baconianism, Tyndall asserts that without the exercise of the imagination, "our knowledge of nature would be a mere tabulation of coexistences and sequences" (1:419). Imagination, not experiment and observation, is "the architect of physical theory," and Tyndall cites the work of Newton, Dalton, Davy, Faraday, and Darwin as examples. By the same token, "elements of scientific inquiry" like "sharpness of observation, . . . detection of analogies apparently remote, . . . [and] classification and organization of facts according to the analogies discerned" are equally part of the "disciplines of the poet." Science, Tyndall concludes, wants to unite with, not divorce from, literature: "For science, however, no exclusive claim is here made; you are not urged to erect it into an idol. . . . The world embraces not only a Newton, but a Shakespeare—not only a Boyle, but a Raphael—not only a Kant, but a Beethoven—not only a Darwin, but a Carlyle. Not in each of these, but in all, is human nature whole. They are not opposed, but supplementary—not mutually exclusive, but reconcilable" (1:453–54, 494).

Tyndall's defense of the imagination is more eloquent and rhapsodic than anything Holmes has to say on the subject. But like science itself, Tyndall's imaginative flight occurs only after establishing the ground from which this flight occurs, and emphasizes the necessity of keeping it under control. New knowledge springs from old knowledge, from the facts that observation and experiment generate—and like Holmes, Tyndall gives only limited credence to the notion that observation and experiment are theory-laden: it is clearly possible to establish some objective facts from which theories are then constructed. Sometimes solutions can be derived mechanically, but more often there will be a need for hypothesis, conjecture. Yet this speculation is not untrammeled: the scientific imagination, based in fact and subject to verification, is a matter not of "guesswork" but of Holmes's "balancing of probabilities." In his Baconian moments Holmes may claim that his success is the result of a method that anyone can, in theory, apply, but Tyndall, like Watson, knows that either there is something more than mere method to Holmes's success, or the method requires some redefinition:

> [T]he investigator proceeds by combining intuition and verification. He ponders the knowledge he possesses, and tries to push it further; he guesses, and checks his guess; he conjectures, and confirms or explodes his conjecture. These guesses and conjectures are by no means leaps in the dark. . . . The force of intellectual penetration . . . is not, as some seem to think, dependent upon method, but upon the genius of the investigator. There is, however, no genius so gifted as not to need control and verification. . . . Thus the vocation of the true experimentalist may be defined as the continued exercise of spiritual insight, and its incessant correction and realization. (1:399)

## THE SCIENCE OF DEDUCTION, NARRATIVE FORM, AND AESTHETICS

Much is made, and rightly so, of the many places where Holmes complains about the way Watson "romanticizes" his stories. Of "A Study in Scarlet," Holmes tells his friend, "'Honestly, I cannot congratulate you upon it. Detection is, or ought to be, an exact science and should be treated in the same cold and unemotional manner. You have attempted to tinge it with romanticism, which produces much the same effect as if you worked a love-story or an elopement into the fifth proposition of Euclid'" (SIGN, 1). According to Holmes, Watson has "degraded what should have been a course of lectures into a series of tales" (COPP); his "fatal habit of looking at everything from the point of view of a story" has "ruined what might have been an instructive and even classical series of demonstrations" (ABBE).

This radical disjunction between lecture and tale, demonstration and story, is belied by Holmes's own sense of himself as an artist, in particular, as an actor with a decided flair for the dramatic. He enjoys closing a case with a well-directed and well-acted coup de theatre. His facility with disguises causes Watson to remark that "the stage lost a fine actor" (SCAN) when he devoted himself to the detection of crime. In "The Adventure of the Dying Detective," where he pretends to be dying of a rare disease to fool both the criminal and Watson, he congratulates himself for carrying out his role "with the thoroughness of the true artist."

"Thoroughness" suggests that Holmes's conception of the "true artist"—like that of the "ideal detective"—is a matter not simply of inspired genius, but of meticulous attention to detail. This notion is echoed in Watson's claim that "Holmes had the impersonal joy of the true artist in his better work" (VALL, 1, 1). Yet "impersonal joy" and "true artist" would seem to fit conflicting models, the impersonal scientist and the personal artist; what are they doing yoked together?

Part of the answer lies in the dramatic aspect of Holmes's artistry. Like an actor, Holmes can completely subsume his own features, bearing, clothes, under those of another man. He loses his own personality and assumes another. But when not engaged in games of counterplotting or disguise, Holmes is still impersonal in that he approaches a case for the sheer intellectual thrill of it; the interest lies in solving a problem. He remains aloof from his clients—they are abstract "units" or "factors." He cares little for money or fame and therefore allows the official police to receive the credit. His investigations often involve crimes of violence or sensitive matters of state, but just as often they do not involve crime at all, but almost mundane mysteries. As long as a case is interesting as a puzzle, he will accept it, though he often complains that with the exception of Professor Moriarity there is virtually no artistry in crime against which to test his artistry in detection: "the days of the great cases are past. Man, or at least criminal man, has lost all enterprise and originality. As to my own practice, it seems to be degenerating into an agency for recovering lost lead pencils and giving advice to young ladies from boarding-schools" (COPP).

This is detection for detection's sake, and it must be linked with the aesthetic movement of the late nineteenth century, not least because both the aesthete and the detective share a tension between the celebration and suppression of personality. The "impersonal joy of the true artist" also finds expression in Wilde's preface to *Dorian Gray*: "To reveal art and conceal the artist is art's aim."[21] Similar views appear in Wilde's essays of this period. In "The Decay of Lying" we are told that "Art never expresses anything but itself"; in "The Critic as Artist," that "Man is least himself when he talks in his own person" and consequently that "[o]ne tires . . . of the work of individuals whose individuality is always noisy."[22] Yet at the same time, "behind everything that is wonderful stands the individual." As he puts it in "The Soul of Man under Socialism," "[a] work of art is the unique result of a unique temperament. . . . Art is the most intense mode of individualism that the world has known." "[I]t is only by intensifying his own personality," Wilde writes in "The Critic as Artist," "that the critic can interpret the personality and work of others." In a typically Wildean paradox, "the objective form is the most subjective in matter" (7:127, 300, 162, 191). The "impersonal" element of Wilde's aesthetic represents a logical extension of the Romantic conception of the artist, whose intense personal expression is channeled into form and pose and hence effaces itself. But the methodology of Wilde's concealed artist stands on intense individuality; Holmes's equally impersonal artistry raises the imagination on a scaffolding of fact. Detection shares with science and art the human and the inhuman, the personal and the impersonal.

It is no wonder that given his own realist aesthetic, Watson bristles at the charge of "romanticizing" his stories. His are, to be sure, "the chronicles of fact" (DANC), but it is precisely because of his commitment to fact that his stories have romantic, even sensational, elements: the cases often *are* romantic or sensational. To the charge that he has "tinged" "A Study in Scarlet" with "romanticism," Watson defends himself by saying that "the romance was there. . . . I couldn't tamper with the facts" (SIGN, 1). The cases, like Holmes's method itself, blend fact and romance inextricably. Newspaper accounts and police reports are the venues where facts are laid out in a lump, yet it is here, Watson says, where "realism [is] pushed to its extreme limits," that the result is "neither fascinating nor artistic" (IDEN). Holmes agrees, arguing in essence that police reports reflect many of the same deficiencies that the police themselves bring to their investigations, namely, attending only to the obvious and failing to identify the clues that are most crucial: "A certain selection and discretion must be used in producing a realistic effect. . . . This is wanting in the police report, where more stress is laid, perhaps, upon the platitudes of the magistrate than upon the details, which to an observer contain the vital essence of the whole matter." Indeed, Watson is later vindicated when Holmes writes one of his own adventures: "I have often had occasion to point out to him [Watson] how superficial are his own accounts and to accuse him of pandering to popular taste instead of confining himself rigidly to facts and figures. 'Try it yourself, Holmes!' he has retorted, and I am compelled to admit that, having taken my pen in my hand, I do begin to realize that the matter must be presented in such a way as may interest the reader" (BLAN).

Realism that is artistic rather than "pushed to extreme limits" is the goal equally of Holmes, Watson, Doyle, and Wilde. For Wilde, writing in "The Decay of Lying," the difference between Zola and Balzac was "the difference between unimaginative realism and imaginative reality" (7:17). Beauty depends on accuracy of detail, but detail selected and shaped. For Doyle, the Holmes stories represented, for all their realism, "that distraction from the worries of life and stimulating change of thought which can only be found in the fairy kingdom of romance."[23] They are "imaginative reality"—or, to invert Wilde's comment about criticism, subjective form is the most objective in matter. The content of the stories is realistic but the effect of the stories is romantic. Such a view is suggestive because it removes various conflicts—not just realism/Romanticism but art/science, intuition/reason, aesthete/ascetic—from within the stories and locates them in the relationship between reader and work. Although Watson's role as narrator and "author" of the stories, as well as the many comments within the stories about the nature of their telling, muddies Doyle's neat distinction, the basic point remains that Doyle does not see his detective as combining, in personality or methods, radically different traditions. Rather, Holmes is both scientist and artist, both logical and intuitive, because science itself brings these together—as, for that matter, does art. Like Tyndall, Holmes is both materialist and celebrator of the scientific mysteries of the imagination; like Wilde, he demands details but rejects Zolaesque narratives. If there be any doubt about the similarity of methods, consider the following in praise of the critic, who for Wilde is the true artist: "Criticism can re-create the past for us from the very smallest fragment of language or art, just as surely as the man of science can from some tiny bone, or the mere impress of a foot upon a rock, re-create for us the winged dragon or Titan lizard that once made the earth shake beneath its tread, can call Behemoth out of his cave, and make Leviathan swim once more across the startled sea" (7:217).

## FROM ANTAGONISM TO DUALITY

It is difficult if not impossible to talk about science and literature without setting up these pairs of opposites. In his fine essay on scientific method and the detective novel, Joseph Agassi works from within the concept of reason to show that reason itself is both mechanical and mysterious, realistic and Romantic. Yet he, too, returns to the comfort of external opposites reconciled internally in detective fiction, which "bridges the gap between the realistic and the romantic; between the cold calculating and the warmhearted intuitive."[24] Even if these pairs are brought together in detective fiction, they are normally irreconcilable in most fiction, most poetry, most science. Ultimately, for Agassi, the "central theme" of Doyle's stories is this "unresolved conflict" between the mechanical and the intuitive.[25]

I have had recourse to such a strategy, and yet I have also tried to emphasize that I employ it for heuristic purposes only, for when we look more closely, the distinctions we impose between these terms begin to break down. Denying this breakdown, turning back to useful but overly simplistic dichotomies, whether our subject is the methodology of Sherlock Holmes or nineteenth-century methodology in general, is not only historically inaccurate; it reinforces the view that some deeply seated disjunction is fundamental or at least inevitable. When we decry the existence of "two cultures," when we applaud those rare fictional bridgings of "gaps," we implicate ourselves in the presence of those very gaps; as students of science and literature, we need rather to show—as recent scholarship seems to be doing—that the concept of schism is not now and never has been sufficiently complex to do justice to the relationship of science and literature. But we must do so not in the spirit of bringing science down a peg: the heyday of positivism is past, and science itself is largely responsible for questioning its adequacy. We must instead find new ways of describing the relationship.

As a tentative suggestion, let me borrow a borrowed term that, in spite of its shortcomings, captures something of what I have in mind: complementarity. In common usage, complementarity describes the coming together of two (or more) things to form a unified whole, each supplying what the other lacks. In physics, complementarity refers to the wave-particle duality of light, the fact that light can be understood only as simultaneously a wave and a particle even though these two states are theoretically incompatible. By forcing us to focus on the integrated whole rather than the individual parts, and through its emphasis on the importance of the human observer, this term is useful for describing the relationship of science and literature.[26]

Complementarity is employed by Caprettini, who uses it to describe the nature of Holmes's thought in "The Adventure of the Lion's Mane." He argues that there are two models of Holmes's mind and methods, the attic and the magnifying glass. Holmes oscillates between these two models both literally and figuratively, for his investigation involves both his usual close observation of the scene of the mystery and his rummaging through his garret of old books in search of a piece of information he can only vaguely remember having once read. Holmes's thought in this story—but the remark can be applied equally well to his actual methods and to all the Holmes canon—"takes place in terms of a complementarity between the *attic* . . . and the *magnifying glass*."[27] For Caprettini, the attic corresponds to the collection and storage of facts, what I have identified as the Baconian elements of Holmes's methods, while the magnifying glass suggests a more directed, theory-laden process of observational inquiry, what I have identified as the more imaginative elements of Holmes's methods and connected with the hypothetico-deductive approach.

Caprettini separates attic and magnifying glass as well as their application; taken together, they show complementarity. Yet I would argue that even taken separately, attic and magnifying glass show within themselves evidence of complementarity. Some selection is involved both in deciding what, originally, to keep in one's brain-attic (the size of which, we know from "A Study in Scarlet," is limited) and, during the case, in determining that a particular piece of information stored in the attic is important enough to be retrieved. Similarly, some objective observation is necessary to establish the basic facts or clues from which selection and theory prepare the way for further observation. There is, in other words, complementarity between attic and magnifying glass, but attic and magnifying glass cannot be understood as mutually exclusive, for they themselves are individually composed of the very complementarity that for Caprettini they achieve when brought together. This layering of complementarity is also what we find as we pass from Doyle to Holmes, on from the scientific detective to Tyndall and Huxley, Cuvier and Zadig, Wilde and Wordsworth, and throughout the nineteenth-century discussion of method. But we must also keep complementarity in mind as we pass forward through our own century, for if the methods of the fictional detective were once best explained by reference to those of the scientist, those of the scientist may now be best explained by reference to those of the fictional detective, as here in Einstein's interpretive appropriation of Holmes:

> In nearly every detective novel since the admirable stories of Conan Doyle there comes a time where the investigator has collected all the facts he needs for at least some phase of his problem. These facts often seem quite strange, incoherent, and wholly unrelated. The great detective, however, realizes that no further investigation is needed at the moment, and that only pure thinking will lead to a correlation of the facts collected. So he plays his violin, or lounges in his armchair enjoying a pipe, when suddenly, by Jove, he has it! Not only does he have an explanation for the clews at hand, but he knows that certain other events must have happened. Since he knows exactly where to look for it, he may go out, if he likes, to collect further confirmation for his theory.
>
> The scientist reading the book of nature, if we may be allowed to repeat that trite phrase, must find the solution himself, for he cannot . . . turn to the end of the book. In our case the reader is also the investigator, seeking to explain, at least in part, the relation of events to their rich context. To obtain even a partial solution the scientist must collect the unordered facts available and make them coherent and understandable by creative thought.[28]

Notes

Index

# Notes

INTRODUCTION

1. *Blackwood's Edinburgh Magazine* 109 (1871): 517–19. "Continuation of an old song" because a similar poem had appeared after the publication of *The Origin of Species.*

2. For the initial articulation of the two cultures, see C. P. Snow, *The Two Cultures and A Second Look*, 2d ed. (Cambridge: Cambridge UP, 1969). The movement away from this manner of approaching the science and literature question is captured by the titles of such recent collections as *One Culture: Essays in Science and Literature*, ed. George Levine (Madison: U of Wisconsin P, 1987) and *Beyond the Two Cultures: Essays on Science, Technology, and Literature*, ed. Joseph W. Slade and Judith Yaross Lee (Ames: Iowa State UP, 1990).

3. George Levine, "The Novel as Scientific Discourse: The Example of Conrad," *Novel* 21 (1988): 223.

4. Edward Dowden, "The Scientific Movement and Literature," *Studies in Literature: 1789–1877*, 6th ed. (London: Kegan Paul, 1892), 85.

CHAPTER 1. THE SCIENCE OF SCIENCE: BACONIAN INDUCTION IN NINETEENTH-CENTURY BRITAIN

1. Abraham Cowley, "To the Royal Society," lines 37–41, 94–99, in Thomas Sprat, *The History of the Royal-Society of London* (London: Martyn, 1667). In Henry Power's 1664 *Experimental Philosophy*, Bacon is called the "Patriarch of Experimental Philosophy" (quoted in Macvey Napier, "Remarks, Illustrative of the Scope and Influence of the Philosophical Writings of Lord Bacon," *Edinburgh Philosophical Transactions* 8 [1818]: 401).

2. Humphry Davy, "Lecture 1—Introductory to Electro-Chemical Science," *The Collected Works of Sir Humphry Davy*, ed. John Davy, 9 vols. (London: Smith, 1839–40), 8:276; "Address of the President on Taking the Chair of the Royal Society, For the First Time . . . ," *Collected Works*, 7:14.

3. John Playfair, *The Works of John Playfair*, ed. James G. Playfair, 4 vols. (Edinburgh: Constable, 1822), 2:115.

4. "Francis Bacon," *Encyclopaedia Britannica*, 9th ed.

5. Herbert W. Blunt, "Bacon's Method of Science," *Proceedings of the Aristotelian Society*, n.s. 4 (1903–4): 24.

6. Henry Hallam, *Introduction to the Literature of Europe in the Fifteenth, Sixteenth, and Seventeenth Centuries*, 5th ed., 4 vols. (London: Murray, 1855), vol. 3, pt. 3, ch. 3, sec. 80.

7. W. Stanley Jevons, *The Principles of Science*, 2d ed. (1877; New York: Dover, 1958), viii. This quote appears in the reprinted preface to the first (1874) edition.

8. Thomas Fowler, introduction, *Bacon's Novum Organum*, 2d ed. (London: Froude, 1888), 135. The bibliography appears on 145–55.

9. John Robertson, introduction, *The Philosophical Works of Francis Bacon*, ed. John Robertson (1905; Freeport, NY: Books for Libraries P, 1970), vii.

10. Susan F. Cannon, *Science in Culture: The Early Victorian Period* (New York: Science History Publications, 1978), 73–74, 228–32.

11. Davy, *Collected Works*, 4:15–16.

12. Another example of both the real existence of a Baconian orthodoxy and the extent of its influence may be found in the *Saturday Review*'s reaction to John Tyndall's 1870 lecture "Scientific Use of the Imagination": what we admire about the scientific mind is the "splendid series of inductions verified step after step by rigorous experiment and observations," the "firmly balanced and duly graduated tread of a mind trained in the discipline of logic, and careful to plant every step on the assured ground of fact or experience," not "the leap of the imagination." Quoted in John Tyndall, *Essays on the Use and Limit of the Imagination in Science*, 2d ed. (London: Longmans, 1870), 2–3.

13. Richard Yeo, "An Idol of the Market-Place: Baconianism in Nineteenth-Century Britain," *History of Science* 23 (1985): 288. Antonio Pérez-Ramos, *Francis Bacon's Idea of Science and the Maker's Knowledge Tradition* (Oxford: Clarendon, 1988), 7–31.

14. Pérez-Ramos, *Francis Bacon's Idea of Science*, 24.

15. Ibid., 19, 26; Yeo,"Idol of the Market-Place," 254–59. For a specific historical example of these associations in which the influence of Coleridge figures prominently, see Adrian Desmond, *The Politics of Evolution: Morphology, Medicine, and Reform in Radical London* (Chicago and London: U of Chicago P, 1989).

16. Yeo, "Idol of the Market-Place," 251–52; Pérez-Ramos, *Francis Bacon's Idea of Science*, 26–30.

17. Samuel Taylor Coleridge, *The Philosophical Lectures of Samuel Taylor Coleridge*, ed. Kathleen Coburn (New York: Philosophical Library, 1949), 333–34.

18. William Whewell, *The Philosophy of the Inductive Sciences*, 2d ed., 2 vols. (1847; New York: Johnson, 1967), vol. 2, bk. 12, ch. 11. Hereafter *Philosophy*.

19. G. H. Lewes, *Comte's Philosophy of the Sciences* (London: Bohn, 1853), pt. 1, sec. 10. Hereafter *Comte's Philosophy*.

20. [Thomas Macaulay], "Lord Bacon," *Edinburgh Review* 65 (1837): 87–88.

21. Ibid., 102. Macaulay was like most of the thinkers discussed in this chapter who saw themselves as Baconians working in or extending the Baconian tradition: when it came to the specifics of the Baconian method, they all struggled to define how they could criticize and revise that tradition while still remaining within it.

22. Modern scholarship has generally supported this view, especially where the connection of Bacon and Newton is concerned. In addition to Yeo on Bacon, for Newton see N. R. Hanson, "Hypotheses Fingo," and L. L. Laudan, "Thomas Reid and the Newtonian Turn of British Methodological Thought," *The Methodological Heritage of Newton*, ed. Robert E. Butts and John W. Davis (Toronto: U of Toronto P, 1970), 17–32, 103–31; and Arnold Thackray, *Atoms and Powers: An Essay on Newtonian Matter-Theory and the Development of Chemistry* (Cambridge: Harvard UP, 1970).

23. Jevons, *Principles of Science*, bk. 4, ch. 23, sec. 1.

24. Auguste Comte, *The Positive Philosophy of Auguste Comte*, freely trans. Harriet Martineau (1853; New York: Blanchard, 1858), 731–32.

25. Kuno Fischer, *Francis Bacon of Verulam*, trans. John Oxenford (London: Longmans, 1857), 124.

26. David Brewster, *Memoirs of the Life, Writings, and Discoveries of Sir Isaac Newton*, 2 vols. (Edinburgh: Constable, 1855), 2:404. I quote from this later, expanded version of Brewster's 1831 *The Life of Sir Isaac Newton*, but the section on Bacon and Newton also appears in the earlier work.

27. Yeo, "Idol of the Market-Place," 267.

28. Augustus De Morgan, *A Budget of Paradoxes*, ed. David Eugene Smith, 2d ed., 2 vols. (1915; Freeport, NY: Books for Libraries P, 1969), 1:75, 84.

29. Jevons, *Principles of Science*, bk. 4, ch. 26, sec. 2.

30. Robert Leslie Ellis, general preface to the philosophical works, *The Works of Francis Bacon*, ed. James Spedding, Robert Leslie Ellis, and Douglas Denon Heath, 14 vols. (London: Longmans, 1857–74), vol. 1, sec. 18. Subsequent references to Ellis's preface cited as "general preface." Since Ellis and Spedding were responsible for all but Bacon's legal writings, this edition is often referred to simply as Ellis and Spedding.

31. Brewster, *Memoirs of the Life*, 2:404.

32. William Whewell, *The History of the Inductive Sciences*, 3 vols. (London: Parker, 1837), vol. 3, bk. 14, ch. 8. Hereafter *History*. Whewell's context is a discussion of mineralogical classification in which he argues that the Germans are far ahead of the English in mineralogy precisely because they are willing to do the theorizing on which systematization depends.

33. [William Whewell], rev. of *The Works of Francis Bacon*, ed. James Spedding, Robert Leslie Ellis, and Douglas Denon Heath, *Edinburgh Review* 106 (1857): 314–15. Hereafter "Review."

34. Whewell, *Philosophy*, vol. 2, bk. 13, ch. 2.

35. *Novum Organum*, bk. 1, aphorism 50: "But by far the greatest hindrance and aberration of the human understanding proceeds from the dulness, incompetency, and deceptions of the senses." (All quotes from Bacon are taken from the English translations in Ellis and Spedding.) Jevons applauds Bacon for this insight (*Principles of Science*, bk. 4, ch. 18, sec. 3).

36. Herbert Spencer, "The Genesis of Science," *Essays: Scientific, Political, and Speculative*, 3 vols. (1891; Osnabrück: Zeller, 1966), 2:3; vols. 13–15 in *The Works of Herbert Spencer.*

37. G. H. Lewes, *The Foundations of a Creed*, 2 vols. (London: Trübner, 1874–75), vol. 1, secs. 14, 61; vols. 1–2 in *Problems of Life and Mind*, 5 vols. (London: Trübner, 1874–79). Hereafter *Foundations.* Nineteenth-century positivism is finally being rescued from the caricatures and distortions of twentieth-century humanists: see Peter Allan Dale, *In Pursuit of a Scientific Culture: Science, Art, and Society in the Victorian Age* (Madison: U of Wisconsin P, 1989), and Edward Davenport, "The Devils of Positivism," *Literature and Science: Theory and Practice*, ed. Stuart Peterfreund (Boston: Northeastern UP, 1990), 17–31.

38. Whewell, *Philosophy*, vol. 1, bk. 1, ch. 2, and vol. 2, bk. 11, chs. 3–5.

39. Ibid., vol. 2, bk. 11, ch. 5. Whewell's emphasis.

40. Ibid., preface, v–x, and vol. 2, bk. 13, ch. 1.

41. Ibid., vol. 1, bk. 1, ch. 1.

42. Comte, *Positive Philosophy*, 27.

43. John Stuart Mill, *A System of Logic Ratiocinative and Inductive*, ed. J. M. Robson, 2 vols. (Toronto: U of Toronto P, 1973–74), bk. 4, ch. 1; vols. 7–8 of *The Collected Works of John Stuart Mill*, gen. ed. J. M. Robson, 33 vols. (Toronto: U of Toronto P, 1962–1991). Hereafter cited by book, chapter, and section as *Logic*.

44. Ellis, general preface, sec. 18.

45. Baron [Justus von] Liebig, "Lord Bacon as Natural Philosopher," *Macmillan's* 8 (1863):241, 263.

46. John F. W. Herschel, *A Preliminary Discourse on the Study of Natural Philosophy* (1830; Chicago: U of Chicago P, 1987), sec. 68.

47. John F. W. Herschel, "Whewell on the Inductive Sciences," *Essays from the Edinburgh and Quarterly Reviews, with Addresses and Other Pieces* (London: Longman, 1857), 244. Hereafter *Essays.*

48. Karl Pearson, *The Grammar of Science*, 3d. ed. (London: Macmillan, 1911), 6. With the exception of two additional chapters, this edition is substantially the same as the first (1892). Pearson's comment is especially striking given that the *Grammar* offers a "sound idealism" in which basic conceptions like cause and effect, matter, force, the simplicity and regularity of nature, etc., are seen as limiting creations of the mind. For a contemporary example of concern for the despair and uncertainty that Pearson could evoke, see the chapter on the *Grammar* in *The Education of Henry Adams.*

49. John Theodore Merz, *A History of European Thought in the Nineteenth Century*, 4 vols. (1896–1914; Edinburgh: Blackwood, 1923), 3:62.

50. Jevons, *Principles of Science*, bk. 4, ch. 18, sec. 3.

51. De Morgan, *Budget of Paradoxes*, 1:80.

52. [Henry Brougham], rev. of *The Bakerian Lecture on the Theory of Light and Colours*, by Thomas Young, *Edinburgh Review* 1 (1803):450.

53. Davy, *Collected Works*, 8:317.

54. Herschel, "An Address to the British Association . . . ," *Essays*, 636. Later in the same address Herschel also claims that "a disposition is becoming prevalent towards lines of speculation, calculated rather to bewilder than

enlighten, . . . by reducing [science] to a mere assemblage of narrow and meaningless facts and laws" (659).

55. Whewell, *Philosophy*, preface, viii, and vol. 2, bk. 12, ch. 11.

56. Whewell, *History*, vol. 1, bk. 5, ch. 2.

57. Herschel, *Preliminary Discourse*, secs. 201, 211–12, 170–71. Cannon suggests that the methodology of the *Preliminary Discourse* represents the more general and orthodox opinion of Herschel's professional group as a whole rather than Herschel's own particular (and less "Baconian") philosophy. See *Science in Culture*, 73–74, and "John Herschel and the Idea of Science," *JHI* 22 (1961):215–39. This may account for some of the methodological tensions in the *Preliminary Discourse*.

58. Brewster, *Memoirs of the Life*, 2:405.

59. John Nichol, *Francis Bacon, His Life and Philosophy*, 2 vols. (Edinburgh: Blackwood, 1907), 2:232.

60. Mill, *Logic*, bk. 3, ch. 8, sec. 3, and bk. 6, ch. 5, sec. 5.

61. Jevons, *Principles of Science*, bk. 4, ch. 19, sec. 1.

62. Ibid., preface, ix.

63. Whewell, *Philosophy*, vol. 2, bk. 12, ch. 11.

64. Ellis, general preface, secs. 4, 5, 9.

65. Whewell, "Review," 313, 316. For a similar assessment in a later, popular medium, see Thomas Fowler, *Bacon* (New York: Putnam, 1881), 141.

66. [Macaulay], "Lord Bacon," 87.

67. Herschel, *Preliminary Discourse*, secs. 22–23.

68. Patrick Brantlinger, introduction, *Energy and Entropy: Science and Culture in Victorian Britain*, ed. Patrick Brantlinger (Bloomington: Indiana UP, 1989), xiii–xxi.

69. Arthur Conan Doyle, "The Adventure of the Blanched Soldier," *The Complete Sherlock Holmes*, 2 vols. (Garden City, NY: Doubleday, n.d.), 2:1011.

70. Gerald Holton, *Thematic Origins of Scientific Thought: Kepler to Einstein*, rev. ed. (Cambridge: Harvard UP, 1988), 35, argues that these translations subtly altered Newton's meaning and thus contributed to the common view that he loathed all hypotheses. Laudan ("Thomas Reid," 108–22) locates the origin of the view of Bacon and Newton's shared hostility to hypotheses in the work of the eighteenth-century Scottish philosopher Thomas Reid.

71. [Brougham], rev. of *Bakerian Lecture*, 451–52.

72. *Novum Organum*, bk. 2, aphorism 20.

73. Brewster, *Memoirs of the Life*, 2:404. Fowler, *Bacon's Novum Organum*, secs. 15–16, and *Bacon*, 120. When the development of thermodynamics enabled admirers of Bacon's method to transform his discussion of heat from a liability into a supporting example, Newtonians countered with arguments that Newton, not Bacon, was the anticipator of the modern theory of heat. See Baden Powell, *History of Natural Philosophy* (London: Longman, 1830), 202; Jevons, *Principles of Science*, bk. 4, ch. 26, sec. 4, John Tyndall, *Heat Considered as a Mode of Motion* (London: Longman, 1863), 25; P. G. Tait, *Lectures on Some Recent*

*Advances in Physical Science,* 2d ed. (London: Macmillan, 1876) 33, 48. For a modern assessment of these claims about the genealogy of thermodynamics, see Yehuda Elkana, *The Discovery of the Conservation of Energy* (Cambridge: Harvard UP, 1974), 57-79.

74. Whewell, *Philosophy,* vol. 1, bk. 5, ch. 1; vol. 2, bk. 11, ch. 5; vol. 2, bk. 12, ch. 13. On Newton's hypotheses, see also *History,* vol. 2. bk. 9, ch. 3.

75. Herschel, *Essays,* 244.

76. Whewell, *Philosophy,* vol. 2, bk. 12, ch. 13. Playfair, *Works,* 2:118.

77. Lewes, *Foundations,* vol. 1, secs. 86–88, 98; *Comte's Philosophy,* pt. 1, sec. 10.

78. Mill, *Logic,* bk. 3, ch. 14, sec. 5.

79. Whewell, *Philosophy,* vol. 2, bk. 12, ch. 13; *History,* vol. 2, bk. 9, ch. 11.

80. Fowler, *Bacon,* 199.

81. Mill, *Logic,* bk. 3, ch. 9, sec. 6.

82. Herschel, *Essays,* 670. For similar assessments, see C. J. Ducasse, "Whewell's Philosophy of Scientific Discovery," *Philosophical Review* 60 (1951): 56–69, 213–34, and E. W. Strong, "William Whewell and John Stuart Mill: Their Controversy about Scientific Knowledge," *JHI* 16 (1955): 209–31. It was Whewell's epistemology, not his views on the mechanics of scientific method, that tended to draw criticism. For a view emphasizing the gap between Mill and Whewell, see Peter Achinstein, *Particles and Waves* (Oxford: Oxford UP, 1991).

83. Comte, *Positive Philosophy,* 199, 799.

84. Mill, *Logic,* bk. 3, ch. 16, sec. 6.

85. Jevons, *Principles of Science,* bk. 4, ch. 23, secs. 1, 3–4; bk. 4, ch. 26, sec. 4.

86. Tyndall, *Essays,* 54.

87. De Morgan, *Budget of Paradoxes,* 1:85–86.

88. Ibid., 1:81.

89. Letter of 20 August 1837. Quoted in David B. Wilson, "Herschel and Whewell's Version of Newtonianism," *JHI* 35 (1974):85n.

90. Whewell, "Review," 316.

91. Herschel, *Preliminary Discourse,* sec. 201.

92. Herschel, *Essays,* 390.

93. Tyndall, *Essays,* 54. That reactions focused on the use rather than the limit of the imagination is shown by the response of the *Times* (19 September 1970) to the second address: "there is, perhaps, more danger of our imagination being exercised too freely than of its not being exercised sufficiently."

94. Liebig, "Lord Bacon as Natural Philosopher," 265.

95. W. B. Carpenter, "Man the Interpreter of Nature," *Victorian Science: A Self-Portrait from the Presidential Addresses of the British Association for the Advancement of Science,* ed. George Basalla, William Coleman, and Robert H. Kargon (Garden City, NY: Doubleday, 1970), 419 (original emphasis). Carpenter's title is taken from the first aphorism of the *Novum Organum.*

96. Comte, *Positive Philosophy,* 452, 795.

97. Lewes, *Comte's Philosophy,* pt. 2, sec. 3.

98. Lewes, *The Life of Goethe,* 2d ed. (1864; London: Dent, 1938), 52–53.

99. Lewes, *Foundations,* vol. 1, sec. 62. For a detailed appraisal of Lewes' evolving position on the relation of science and art, see Peter Allan Dale, "George Lewes's Scientific Aesthetic: Restructuring the Ideology of Symbol," *One Culture: Essays in Science and Literature,* ed. George Levine (Madison: U of Wisconsin P, 1987), 92–116. Dale argues that Lewes' changing aesthetic theory alters his views of scientific method.

100. William Wordsworth, *The Prose Works of William Wordsworth,* ed. W. J. B. Owen and Jane Worthington Smyser, 3 vols. (Oxford: Clarendon, 1974), 1:141.

101. Pearson, *Grammar of Science,* 35.

102. John Tyndall, "Scientific Use of the Imagination," *Fragments of Science,* 6th ed., 2 vols. (New York: Burt, n.d.), 1:494. This passage does not appear in the earliest versions of the address.

103. For similar assessments of Victorian scientists in different contexts, see Tess Cosslett, *The "Scientific Movement" and Victorian Literature* (Brighton: Harvester, 1982), passim.; W. David Shaw, "Poetic Truth in a Scientific Age," *Centre and Labyrinth: Essays in Honour of Northrop Frye,* ed. Eleanor Cook et al. (Toronto U of Toronto P, 1983), 245–63. Donald R. Benson, "Facts and Constructs: Victorian Humanists and Scientific Theorists on Scientific Knowledge," *Victorian Science and Victorian Values: Literary Perspectives,* ed. James Paradis and Thomas Postlewait (New Brunswick, NJ: Rutgers UP, 1985), 299–318, argues that leading Victorian "Humanists" (especially Pater, Arnold, Ruskin) maintain a basically naive Baconian view of science even though the "Scientific Theorists" (Jevons, Pearson, Tyndall, Huxley) were challenging that very view. I am of course in sympathy with Benson's efforts, but I believe that the rejection of naive Baconianism is much more widespread and begins much earlier than he is aware of, and that he thus underestimates the humanists' involvement in this rejection, seeing nothing but opposition where there is in fact considerable congruence.

104. For a brief discussion of this ambivalence among the major figures in the debate, including Whewell, see Yeo, "Idol of the Market-Place," 251–52, 268–69, 271, 274, 277.

105. Whewell, *Philosophy,* dedication, iv; preface, vi. Whewell's emphasis.

106. Ibid., preface, v–vi.

107. The Inductive Tables are discussed in *Philosophy,* vol. 2, bk. 11, ch. 6.

108. Whewell discusses Cowley's ode in *Philosophy,* vol. 2, bk. 12, ch. 11, mentioning both the Moses image and the reference to Bacon as lord chancellor of both king and nature. Whewell neither endorses nor repudiates these images in his discussion, but his comment that the "fertility and ingenuity of comparison which characterize Cowley's poetry . . . are in this instance largely employed for the embellishment of his subject" suggests that he considered Cowley's comparisons exaggerated.

109. Whewell, *Philosophy,* vol. 2, bk. 12, ch. 11.

110. For Bacon as hero, with special reference to the role of spokesmen for the "new philosophy" like Sprat, see John M. Steadman, "Beyond Hercules: Bacon and the Scientist as Hero," *Studies in the Literary Imagination* 4 (1971): 3–47. For a discussion of Cowley's ode that focuses on aspects other than the Moses image in the creation of a Baconian myth, see Achsah Guibbory, "Imitation and Originality: Cowley and Bacon's Vision of Progress," *SEL* 29 (1989): 99–120.

111. Powell, *History of Natural Philosophy,* 211. Powell ignores Bacon the lawgiver because, in spite of all his praise for Bacon, he cannot see much of practical value in the details of Bacon's method.

112. Hallam, *Introduction to the Literature of Europe,* vol. 3, pt. 3, ch. 3, sec. 80.

113. Bacon often makes this comparison: the frontispiece of the *Novum Organum* shows a ship sailing through the Pillars of Hercules. See Steadman, "Beyond Hercules," 35–38.

114. Herschel, *Essays,* 258–60.

115. Playfair, *Works,* 2:69–70.

116. [Macaulay], "Lord Bacon," 102–3.

117. G. H. Lewes, *A Biographical History of Philosophy* (1845; London: Routledge, 1900), 361, 377–78.

118. Ellis, general preface, sec. 18.

119. Fowler, *Bacon's Novum Organum,* 130; Nichol, *Francis Bacon,* 2:237, 243.

120. Liebig, "Lord Bacon as Natural Philosopher," 242.

121. Merz, *History of European Thought,* 1:95–97.

122. DeMorgan, *Budget of Paradoxes,* 1:84n.

123. Playfair, *Works,* 2:118. T. H. Huxley, "On Descartes' 'Discourse Touching the Method of Using One's Reason Rightly and of Seeking Scientific Truth,'" *Lay Sermons, Addresses, and Reviews* (New York: Appleton, 1870), 342.

124. *Punch* 69 (1871):145.

125. Charles Darwin, *The Autobiography of Charles Darwin,* ed. Nora Barlow (1958; London: Pickering, 1989), 144; vol. 29 in *The Works of Charles Darwin,* ed. Paul H. Barrett and R. B. Freeman, 29 vols. (London: Pickering, 1986–89). T. H. Huxley, "Darwin on the Origin of Species," *Westminster Review,* n.s. 17 (1860):566–68; G. H. Lewes, "Mr. Darwin's Hypotheses," *Fortnightly Review* 9 (1868):355, 370; Mill, *Logic,* bk. 3, ch. 14, sec. 5; Pearson, *Grammar of Science,* 32.

126. See the essays by Roy MacLeod, A. D. Orange, and Richard Yeo in *The Parliament of Science,* ed. Roy MacLeod and Peter Collins (Northwood: Science Reviews, 1981), and Jack Morrell and Arnold Thackray, *Gentlemen of Science* (Oxford: Clarendon, 1981). For the enunciation of the initial Baconian program of the association, see the text of W. Vernon Harcourt's address at the inaugural meeting, *Victorian Science: A Self-Portrait,* 29–44.

127. Prince Albert, "Science and the State," *Victorian Science: A Self-Portrait,* 56 (original emphasis).

CHAPTER 2. REWEAVING THE RAINBOW: ROMANTIC METHODOLOGIES OF POETRY AND SCIENCE

1. George Dodd, "Wordsworth and Hamilton," *Nature* 228 (1970): 1261.

2. The motto (from "Miscellaneous Sonnets 34") appears on the masthead at least through 1935 and on the title page of the bound volumes through 1963. Articles about Wordsworth and science appear throughout *Nature*'s history, occasionally generating heated discussion in the letters to the editor.

3. Some of the general discussions which subvert this supposed antithesis include A. N. Whitehead, *Science and the Modern World* (1925; New York: Free Press, 1967), 82–88; M. H. Abrams, *The Mirror and the Lamp* (New York: Oxford UP, 1953), 308–20; D. M. Knight, "The Physical Sciences and the Romantic Movement," *History of Science* 9 (1970): 54–76; Edward Proffitt, "Science and Romanticism," *Georgia Review* 34 (1980): 55–80; Hans Eichner, "The Rise of Modern Science and the Genesis of Romanticism," *PMLA* 97 (1982): 8–30. Wordsworth's attitude toward science has been an issue of almost continual discussion. Coleridge's science and its place in his philosophy has received much attention in recent years; two detailed studies are Trevor H. Levere, *Poetry Realized in Nature: Samuel Taylor Coleridge and Early Nineteenth-Century Science* (Cambridge: Cambridge UP, 1981) and Raimonda Modiano, *Coleridge and the Concept of Nature* (Tallahassee: Florida State UP, 1985).

4. Charles L. Pittman, "A Biographical Note on Wordsworth," *Bulletin of Furman University* 29.3 (1946): 31–53.

5. "Lamia," 2.229–37. For an account of the significance of the Romantic rejection of the Newtonian explanation of the rainbow, see Abrams, *Mirror and the Lamp*, 308–9.

6. Quoted in Levere, *Poetry Realized in Nature*, 153.

7. Joel Black, introduction, "Newtonian Mechanics and the Romantic Rebellion," *Beyond the Two Cultures: Essays on Science, Technology, and Literature*, ed. Joseph W. Slade and Judith Yaross Lee (Ames: Iowa State UP, 1990), 133.

8. William Wordsworth, *The Prose Works of William Wordsworth*, ed. W. J. B. Owen and Jane Worthington Smyser, 3 vols. (Oxford: Clarendon, 1974), 1:141. This edition contains the 1850 version of the preface, but all versions after 1802 received only minor alterations. Subsequent references to Wordsworth's prose are from this edition and are noted parenthetically in the text.

9. James H. Averill, "Wordsworth and 'Natural Science': The Poetry of 1798," *Journal of English and Germanic Philology* 77 (1978): 232–46, shows that the ambiguity about science in the preface is clarified in the language of the poems themselves. Averill argues that Wordsworth employs the eighteenth-century language of biological classification and chemical analysis to create "a good science and an invidious one" (235). Not only does Wordsworth's interest in science during this period give the *Lyrical Ballads* "a scientific tinge" (238), but the poems themselves "make a place for science in the process of Imagination" (236).

10. Abrams, *Mirror and the Lamp*, 101, 310.

11. Leigh Hunt, "What Is Poetry?" *Imagination and Fancy* (1844; London: Smith, Elder, 1891), 3–4.

12. Ibid., 4.

13. Thomas De Quincey, "Letters to a Young Man Whose Education has been Neglected," *The Collected Writings of Thomas De Quincey*, ed. David Masson, 14 vols. (Edinburgh: Black, 1890), 10:48n. Subsequent references to De Quincey's *Collected Writings* are noted parenthetically.

14. Abrams, *Mirror and the Lamp*, 103–4.

15. Coleridge to Davy, 9 October 1800. Samuel Taylor Coleridge, *Collected Letters of Samuel Taylor Coleridge*, ed. E. L. Griggs, 6 vols. (Oxford: Clarendon, 1956–71), 1:631.

16. William Wordsworth, "Note to 'The Thorn,'" *Literary Criticism of William Wordsworth*, ed. Paul M. Zaul (Lincoln: U of Nebraska P, 1966), 13.

17. Humphry Davy, *The Collected Works of Sir Humphry Davy*, ed. John Davy, 9 vols. (London: Smith, 1839–40), 8:167, 177. Subsequent references will be noted parenthetically in the text.

18. Roger Sharrock, "The Chemist and the Poet: Sir Humphry Davy and the Preface to the *Lyrical Ballads*," *Notes and Records of the Royal Society* 17 (1962): 57–76. A similar argument is offered in Sharrock's "Wordsworth on Science and Poetry," *REL* 3.4 (1962): 42–50. Sharrock's views center on verbal connections between Davy's discourse and Wordsworth's preface, and on the relative status of the two disciplines, rather than on the methodological connections I am highlighting. Sharrock considers the passage from the 1802 preface to be more favorable to the scientist than any other pronouncement by Wordsworth or Coleridge—a view that cannot be maintained, especially for Coleridge—but I should point out that the quotes I have used from the preface are not exclusively from the section added in 1802. Wordsworth's scientific language and modified inductivism are not, therefore, merely a response to Davy; they are, rather, partially in place even before Davy's lecture.

19. W. J. B. Owen, *Wordsworth as Critic* (Toronto: U of Toronto P, 1969), 151–87.

20. Ibid., 155.

21. Michel Foucault, *The Order of Things: An Archaeology of the Human Sciences* (1971; New York: Vintage-Random, 1973), ch. 3, 5, 7, and 8.

22. Ibid., 131.

23. Here again, as Abrams (*Mirror and the Lamp*, 53) notes, later Romantics often saw this "things as they are"/"things as they appear" distinction as the distinction between science and poetry. Yet Wordsworth claims that "things as they are" is bad science, and that "genuine poetry" and "pure science" need not be separated in this way, although he does not specify here how science is to pursue this phenomenological line.

24. This comment appears in the Fenwick Notes to "This Lawn, a carpet all alive," *The Poetical Works of William Wordsworth*, ed. Ernest de Selincourt and Helen Darbishire, 5 vols. (Oxford: Clarendon, 1940–49), 4:425.

25. Quoted in Edith C. Batho, *The Later Wordsworth* (1934; New York: Russell, 1963), 30.

26. Samuel Taylor Coleridge, *The Friend*, ed. Barbara E. Rooke, 2 vols. (London: Routledge; Princeton: Princeton UP, 1969), 1:446, 449; vol. 4 of *The Collected Works of Samuel Taylor Coleridge*, gen. ed. Kathleen Coburn, 11 vols. to date (London: Routledge; Princeton: Princeton UP, 1969– ). Subsequent references will appear parenthetically in the text. *The Friend* originally appeared in irregular weekly intervals in 1809–10, but Coleridge published a three-volume edition, much revised and with considerable new material, including the "Essay on Method," in 1818.

27. Coleridge to Derwent Coleridge, November 1818, *Collected Letters*, 4:885–86. See also Coleridge to J. Britton, 28 February 1819, *Collected Letters*, 4:925, where he ranks the third volume of *The Friend* with the second volume of the *Biographia* and a few of his poems.

28. Samuel Taylor Coleridge, *The Philosophical Lectures of Samuel Taylor Coleridge*, ed. Kathleen Coburn (New York: Philosophical Library, 1949), 334.

29. Richard Yeo, "Reading Encyclopedias: Science and the Organization of Knowledge in British Dictionaries of Arts and Sciences, 1730–1850," *Isis* 82 (1991): 24–49, especially 34–43. See also Alice Snyder's introduction to her *S. T. Coleridge's Treatise on Method* (London: Constable, 1934).

30. For a fuller discussion of this distinction and its place in Coleridge, see Owen Barfield, *What Coleridge Thought* (Middletown, CT: Wesleyan UP, 1971), 23–40.

31. In a note made to another manuscript copy of *The Friend*, Coleridge writes that "Idea and Law are the Subjective and Objective *Poles* of the same Magnet—i.e. of the same living and energizing Reason. What is Idea in the Subject, i.e. in the Mind, is a Law in the Object, i.e. in Nature" (1:497n).

32. *Philosophical Lectures*, 357. Baconian induction, in other words, is wrongly understood as replacing deductive reasoning with the gradual movement from particulars to universals. Ungrounded speculation is not methodical; generalization—not simply from a string of particulars, but with a mental initiative included—is the hallmark of method. The inductive and deductive tendencies, whether in Bacon or Newton or Shakespeare, are thus not separate or opposed, but interconnected and balanced.

33. The importance of the method of the Fine Arts was diminished in the published *Metropolitana*, where it ceased to be an exemplar of *all* method. See Snyder, *S. T. Colleridge's Treatise on Method*, xix–xxii, 25–26, 62.

34. In the *Biographia*, the Imagination is said to be both active and passive and therefore the mind's "intermediate faculty" (*Collected Works*, 1:124). My discussion here builds on Jonathan Arac's exposure of New Criticism's misleading elevation of the Imagination at the expense of the Reason in *Critical Genealogies: Historical Situations for Postmodern Literary Studies* (New York: Columbia UP, 1987), 83–95. Arac's suggestive comment that New Criticism's simultaneous banishment of Shelley (who also tried to reconcile Bacon and Plato) may have had its roots in Shelley's radical political and religious views reminds us that Coleridge's discussion of method is partly motivated by an

effort to distance Baconianism from the democratic and atheistic associations given to it by its French appropriators.

35. Arthur O. Lovejoy, *The Reason, the Understanding, and Time* (Baltimore: Johns Hopkins UP, 1961), 27–28.

36. Interestingly, in his manuscript *Logic,* Coleridge complains that matter of fact is "a phrase . . . that has been vulgarized by its vulgar use, but [which,] in its origin and literal import, is strictly and beautifully expressive." Coleridge considers "matter of fact" used as "evidence of the senses" to be a "vulgar" definition, and indeed, the *OED* shows that "matter of fact" had traditionally been used to signify the difference between established truth and mere opinion, but that other coinages (e.g., "matter-of-factism" and " matter-of-factists") were specifically applied later in the century to positivism. Elsewhere, Coleridge is content to use it in the more vulgar fashion, although he is careful to distinguish between matter of fact and what he considers science. See Samuel Taylor Coleridge, *Logic,* ed. J. R. de J. Jackson (Princeton: Princeton UP; London: Routledge, 1981), 36; vol. 13 of *Collected Works* (subsequent references in parentheses in the text). The *Logic* is a difficult work to date, but much of it seems to be a product of the same period as the *Biographia,* the 1818 *Friend,* and the philosophical lectures.

37. *Coleridge on the Seventeenth Century,* ed. Roberta Florence Brinkley (Durham: Duke UP, 1955), 58.

38. Coleridge to William Godwin, 4 June 1803, *Collected Letters,* 2:947.

39. *Coleridge on the Seventeenth Century,* 41–42.

40. Coleridge to Edward Coleridge, 7 September 1825, *Collected Letters,* 5:492–93. In a letter to Montagu on 1 May 1827 (*Collected Letters,* 6:676), Coleridge says he has enough scattered notes on Bacon to refute Goethe's critical assessment of Bacon's life and character as well as to write a full comparison of Bacon and Plato. Best remembered for his precipitation of the quarrel between Wordsworth and Coleridge in 1810, Montagu knew Coleridge for some forty years. Coleridge invoked Montagu's assistance in dealings with London publishers, especially over the initial periodical publication of *The Friend* between 1808 and 1810. Montagu's sixteen-volume edition of Bacon appeared between 1825 and 1836; in the preface to the work he lauds Coleridge as his "friend and instructor" and acknowledges Coleridge's support and assistance.

41. Snyder, *S. T. Coleridge's Treatise on Method,* 40.

42. Philosophical Lectures, 333.

43. Cf. *Biographia,* 1:255–57.

44. Coleridge to John Taylor Coleridge, 8 April 1825, *Collected Letters,* 5:421.

45. Coleridge to Thomas Poole, 23 March 1801, *Collected Letters,* 2:709.

46. *The Notebooks of Samuel Taylor Coleridge,* ed. Kathleen Coburn, 4 vols. to date (Princeton: Princeton UP, 1957– ), vol. 1, pt. 2, 459.

47. Humphry Davy, *Fragmentary Remains, Literary and Scientific,* ed. John Davy (London: Churchill, 1858), 315. Subsequent references are cited parenthetically in the text as *Remains.*

48. Foucault, *Order of Things*, 218.

49. [Henry Brougham], rev. of *The Bakerian Lecture on Some Chemical Agencies of Electricity*, by Humphry Davy, *Edinburgh Review* 11 (1807–8): 398. Brougham's next review of Davy's work (*Edinburgh Review* 12 (1808): 394–401) opens with the comments about Davy's luck and lack of genius that enraged Coleridge.

50. Coleridge to Davy, 7 December 1808, *Collected Letters*, 3:135–36.

## CHAPTER 3. SEEING THROUGH LYELL'S EYES: THE UNIFORMITARIAN IMAGINATION AND *THE VOYAGE OF THE BEAGLE*

1. John F. W. Herschel, *A Preliminary Discourse on the Study of Natural Philosophy* (1830; Chicago: U of Chicago P, 1987), sec. 65.

2. Charles Lyell, *The Principles of Geology*, 3 vols. (London: Murray, 1830–33), 1:78–79. Hereafter *Principles*.

3. Charles Darwin, *The Autobiography of Charles Darwin*, ed. Nora Barlow (1958; London: Pickering, 1989), 129, 107–9; vol. 29 of *The Works of Charles Darwin*, ed. Paul H. Barrett and R. B. Freeman, 29 vols. (London: Pickering, 1986–89). Hereafter *Works*.

4. Darwin, *Autobiography*, 129.

5. Howard E. Gruber and Valmai Gruber, "The Eye of Reason: Darwin's Development during the *Beagle* Voyage," *Isis* 53 (1962): 186–200. For the published comments about Cape Verde geology, see *Journal of Researches into the Geology and Natural History of the Various Countries Visited by the H.M.S. Beagle*, 1st ed., 2 vols. (1839; London: Pickering, 1986), 1:8; vols. 2 and 3 of *Works*. Hereafter *Journal*. Subsequent references are from this edition unless otherwise noted. References from the second edition are found in Charles Darwin, *The Voyage of the Beagle*, ed. Leonard Engel (New York: Anchor-Doubleday, 1962).

6. Charles Darwin, *The Correspondence of Charles Darwin*, ed. Frederick Burkhardt and Sydney Smith, 8 vols. to date (Cambridge: Cambridge UP, 1985- ), 1:460. Hereafter cited as *Correspondence*.

7. Darwin, *Correspondence*, 1:232, 237, 238, 397. See also Sandra Herbert, "Darwin the Young Geologist," *The Darwinian Heritage*, ed. David Kohn (Princeton: Princeton UP, 1985), 483–510.

8. Charles Lyell, *The Life, Letters and Journals of Sir Charles Lyell*, ed. Mrs. Lyell, 2 vols. (London: Murray, 1881), 2:6, 40.

9. Quoted in Paul H. Barrett and R. B. Freeman, introduction, *Diary of the Voyage of the H.M.S. Beagle*, by Charles Darwin, ed. Nora Barlow (1933; London: Pickering, 1986), xxix; vol. 1 of *Works*. Hereafter *Diary*. See also Gruber and Gruber, "Eye of Reason," 188–89. The first edition is *Journal of Researches into the Geology and Natural History . . .* , but the second edition reverses the order of the two fields.

10. Darwin, *Correspondence*, 2:104.

11. Charles Darwin, *The Life and Letters of Charles Darwin*, ed. Francis Darwin, 2 vols. (New York: Appleton, 1888), 2:55.

12. Martin J. S. Rudwick, "The Strategy of Lyell's *Principles of Geology*," *Isis* 61 (1970): 11; Stephen Jay Gould, *Time's Arrow, Time's Cycle: Myth and Metaphor in the Discovery of Geological Time* (Cambridge: Harvard UP, 1987), 23.

13. For an account of the relationship between natural theology and geology, see C. C. Gillispie, *Genesis and Geology* (Cambridge: Harvard UP, 1951). For historical information see Mott T. Greene, *Geology in the Nineteenth Century* (Ithaca: Cornell UP, 1982), and Walter F. Cannon, "The Uniformitarian-Catastrophist Debate," *Isis* 51 (1960): 38–55.

14. Much recent scholarship in the history of nineteenth-century geology has sought to correct what is now being seen as an overemphasis on Lyell and uniformitarianism; however, as James Secord has noted in his *Controversy in Victorian Geology: The Cambrian-Silurian Dispute* (Princeton: Princeton UP, 1986), such highly theoretical questions as "the reality of a recent flood, the uniformity of natural processes, the age of the world, the progressive history of life, and the origin of man," while not necessarily matters of everyday professional concern for geologists, were "of profound importance for the intellectual life of the nineteenth century" and therefore especially prominent in "popular reviews, prefaces, and presidential addresses" (24).

15. Gould, *Time's Arrow*, 105. Two points are crucial here: first, volcanic eruptions, earthquakes, and floods, though "catastrophic" in comparison with processes like erosion, are not catastrophic to the uniformitarian. Second, catastrophism is not incompatible with the action of small forces over long periods of time; what catastrophists rejected were Lyell's claims that past forces have always been like those of the present (and thus that studying the present can provide a complete explanation of the past) and the notion that the earth's history is not progressive.

16. Ibid., 150–67.

17. Lyell, *Life, Letters and Journals*, 1:2–3.

18. This account is also Lyell's in *Principles*, 1:71–72. See also Roy Porter, *The Making of Geology: Earth Science in Britain*, 1660–1815 (Cambridge: Cambridge UP, 1977), 146–56, 184–204.

19. See Greene, *Geology in the Nineteenth Century* 19–53; Gould, *Time's Arrow*, passim.; and Gillispie, *Genesis and Geology*, chs. 2–5.

20. Gould 143.

21. Lyell, "Scientific Institutions," *Quarterly Review* 34 (1826): 163.

22. Lyell, rev. of *Transactions of the Geological Society*, *Quarterly Review* 34 (1826): 507; "Scrope's *Geology of Central France*," *Quarterly Review* 36 (1828): 441.

23. Lyell, "Scrope's *Geology*," 473.

24. Ibid., 468–69.

25. Herschel, *Preliminary Discourse*, secs. 314, 320.

26. *Table Talk* of 29 June 1833, *The Table Talk and Omniana of Samuel Taylor Coleridge*, ed. T. Ashe (London: Bell, 1888), 230–31.

27. [William Whewell], "Lyell's *Geology* (vol. II)—Changes in the Organic World Now in Progress," *Quarterly Review* 47 (1832): 126; *History of the Inductive*

*Sciences*, vol. 3, bk. 18, ch. 7; *Philosophy of the Inductive Sciences*, vol. 1, bk. 10, ch. 2–3; vol. 2, bk. 13, ch. 8. For a thorough discussion of the reactions of Herschel and Whewell to Lyell's theory, see Michael Ruse, "Charles Lyell and the Philosophers of Science," *BJHS* 9 (1976): 121–31.

28. Both terms are common in geological and historical discourse. Herschel calls fossils "wrecks of a former state of nature," "wonderfully preserved" like "ancient medals and inscriptions in the ruins of an empire" (*Preliminary Discourse*, sec. 317). William Buckland, *Geology and Mineralogy Considered with Reference to Natural Theology*, 2 vols. (London: Pickering, 1836), 1:7, says that geology enables us "to extract from the archives of the interior of the earth, intelligible records of former conditions of our planet, and to decipher documents, which were a sealed book to our predecessors in the attempt to illustrate subterranean history." Lyell's *Manual of Elementary Geology* was subtitled "The Ancient Changes of the Earth and Its Inhabitants as Illustrated by Geological Monuments."

29. See also Rudwick, "Strategy of Lyell's *Principles of Geology*," 4, 16.

30. This is a two-way street, for the historians of the period openly avowed their search for "scientific laws" in history, and often used scientific metaphors to describe their endeavors. For a detailed study of the science of the major Victorian historians, see Rosemary Jann, *The Art and Science of Victorian History* (Columbus: Ohio State UP, 1985).

31. Charles Lyell, *Elements of Geology* (London: Murray, 1838), 1.

32. Niebuhr's work was available in English several years before the publication of the *Principles*, but his ideas were already influential at Oxford and Cambridge in the early 1820s. For information on Niebuhr and his historiography, see Susan Cannon, *Science in Culture: The Early Victorian Period* (New York: Science History Publications, 1978), 46–54; Jann, *Art and Science of Victorian History*, xxii–xxiv; Robert D. Preyer, "The Romantic Tide Reaches Trinity," *Victorian Science and Victorian Values: Literary Perspectives*, ed. James Paradis and Thomas Postlewait (New Brunswick, NJ: Rutgers UP, 1985), 43–47; and Duncan Forbes, *The Liberal Anglican Idea of History* (Cambridge: Cambridge UP, 1952), 12–19. My methodological comparison of Lyell and Niebuhr relies heavily on Roy Porter, "Charles Lyell and the Principles of the History of Geology," *BJHS* 9 (1976): 91–103.

33. Darwin, *Autobiography*, 142.

34. See, for example, *Journal*, 1:80, 157–166, 170–71, and 2:246, 303, 428–29.

35. Generally speaking, Darwin's uniformitarian imagination is cultivated more extensively in the successive texts of the *Diary* and the first and second editions of the *Journal*. A striking example of this is in Darwin's treatment of social and political events. The *Diary* contains numerous and often detailed accounts of such events as the debate over the Reform Bill, political turmoil in South America, Fitzroy's vicious punishments of drunken sailors, and deaths of *Beagle* crew members. Most of these accounts, smacking of catastrophe and revolution, were excised completely from the published version; those that

are mentioned are usually treated briefly and critically. The *Beagle*'s involvement in suppressing a mutiny in Montevideo, for example, dominates an almost three-week span in Darwin's *Diary*—near the outset of the crisis he writes that "we certainly are a most unquiet ship; peace flies before our steps" (78), and at its height he speaks of "an eventful day in the history of the Beagle" (80)—yet it is not mentioned in the published *Journal*. Similarly, the revolution into which he stumbled near Buenos Aires is spoken of with disparagement and annoyance. Such a strategy is in keeping with Darwin's efforts to construct an evolutionary theory that could not be yoked with Radical politics. See Adrian Desmond and James Moore, *Darwin: The Life of a Tormented Evolutionist* (New York: Warner Books, 1991).

36. James Paradis, "Darwin and Landscape," *Victorian Science and Victorian Values*, 87–88, 95–99. Paradis' position assumes a fundamental disjunction in the Romantic mind between poet and scientist, and connects this disjunction to Wordsworth in the *Lyrical Ballads* and Coleridge in *The Friend*; according to my reading of Wordsworth and Coleridge in Chapter 2, such a view has serious problems.

37. Stanley Edgar Hyman, "The Whole World Round," *NMQ* 29 (1959): 277.

38. I am indebted to Jonathan Arac for suggesting this comparison.

39. Lest we think that Darwin couldn't possibly have had such considerations in mind, notice what he deletes after these lines. In the passage quoted in the *Journal*, the wilderness's mysterious tongue teaches awful doubt. But Shelley's poem does not stop there; it offers an alternative to "awful doubt": "or faith so mild, / So solemn, so serene, that man may be / But for such faith with nature reconciled" (lines 77–79). Is the Darwin of 1845, after the reading of Malthus, reluctant to acknowledge that the mysterious tongue of nature teaches faith and reconciliation?

40. Paradis, "Darwin and Landscape," 99.

41. Charles Darwin, *Charles Darwin's Notebooks, 1836-1844*, ed. Paul H. Barrett et al. (Cambridge: Cambridge UP, 1987), 529.

42. Edward Manier, *The Young Darwin and His Cultural Circle* (Dordrecht: Reidel, 1978), 89–96, explores Wordsworth's possible influence on Darwin during this period, but his focus is on the early drafts of Darwin's evolutionary theory and on *The Excursion*, which Darwin said he had read twice through. Although he says the evidence is circumstantial, Manier concludes that Darwin, recently returned from his own excursion, would have been sympathetic to the critique of science in Wordsworth's poem. Manier does not note that this reference is to the preface to *Lyrical Ballads*, but it obviously supports his claim.

43. Darwin, Autobiography, 107.

44. Cannon, *Science in Culture*, 80–96.

45. Alexander von Humboldt, *Personal Narrative . . .* , trans. Helen Maria Williams, 7 vols. (London: Longmans, 1814–29), 1:xxxviii.

46. Ibid., 1:xlii.

47. Darwin, *Correspondence*, 1:503.

48. Ibid., 2:11.

49. For a thorough discussion of Darwin's chronological distortions, see John Tallmadge, "From Chronicle to Quest: The Shaping of Darwin's 'Voyage of the *Beagle*,'" *VS* 23 (1980): 331–33.

50. Ibid., 333.

51. Ibid.

CHAPTER 4. THE "WONDERFUL GEOLOGICAL STORY": UNIFORMITARIANISM AND *THE MILL ON THE FLOSS*

1. Barbara Hardy, *The Novels of George Eliot* (London: Athlone, 1959), 57; *Particularities* (Athens: Ohio UP, 1983), 63.

2. Sally Shuttleworth, *George Eliot and Nineteenth-Century Science* (Cambridge: Cambridge UP, 1984), 63. Subsequent references are cited in the text.

3. Most commentators have either ignored the novel's geology or subsumed it in passing references to Darwinism. Although Rosemary Ashton treats the novel's natural historical allusions more thoroughly in *The Mill on the Floss: A Natural History* (Boston: Twayne, 1990), 30–41, she, too, gives no separate attention to geology. Even Shuttleworth, who does attend to geological concerns in her excellent study, fails to make some of these distinctions. Although I will be arguing against Shuttleworth's views on uniformitarianism, I am indebted to her reading of the Darwinian elements of the novel.

4. George Eliot, *The George Eliot Letters*, ed. Gordon S. Haight, 9 vols. (New Haven: Yale UP, 1954-78), 3:227. Hereafter cited in the text as *Letters*.

5. For a full discussion of the different uniformities, see Stephen Jay Gould, *Time's Arrow, Time's Cycle: Myth and Metaphor in the Discovery of Geological Time* (Cambridge: Harvard UP, 1987).

6. Thomas Pinney, ed., *The Essays of George Eliot* (New York: Columbia UP, 1963), 31n. Hereafter cited in the text as *Essays*.

7. Martin J. S. Rudwick's *Great Devonian Controversy* (Chicago: U of Chicago P, 1985) shows how that important geological debate produced a consensus that changes in organic life have been steady and piecemeal, as Lyell claimed, rather than catastrophic (see especially 355–56, 372–73, 387, 390, 415). The controversy was debated and resolved during the initial years of Eliot's interest in geology.

8. Charles Lyell, *The Geological Evidences of the Antiquity of Man*, 3d ed. (London: Murray, 1863), 412. Lyell's conversion was not registered in the *Principles* until the tenth (1866) edition.

9. Herbert Spencer, "Illogical Geology," *Essays: Scientific, Political, and Speculative*, 3 vols. (Osnabrück: Zeller, 1966), 1:226, 227n.

10. For the various readings that Darwin's language permits, see Gillian Beer, *Darwin's Plots: Evolutionary Narrative in Darwin, George Eliot, and Nineteenth-Century Fiction* (1983; London: Routledge-Ark, 1985), 123–45.

11. Gordon S. Haight, *George Eliot: A Biography* (New York: Oxford UP, 1968), 36.

12. William Buckland, *Geology and Mineralogy Considered with Reference to Natural Theology,* 2 vols. (London: Pickering, 1836), 1:7, 95n. Subsequent references are noted parenthetically.

13. For a critical assessment of the treatises, see C. C. Gillispie, *Genesis and Geology* (Cambridge: Harvard UP, 1951), 209–16. For a more sympathetic account of the factors producing all the Treatises, but especially Buckland's, see Nicolaas A. Rupke, *The Great Chain of History: William Buckland and the English School of Geology* (Oxford: Clarendon, 1983), 193–94, 215–18, 233–54. Rupke argues that Buckland was not a catastrophist, but that is true only in the sense that he was not the caricature catastrophist of Lyell's rhetorical excesses.

14. Lyell was not openly hostile to natural theology and sometimes employed its language, as Darwin did in the *Origin*. More often, however, Lyell simply avoids reference to the arguments of natural theology. This silence was Smith's chief complaint against the *Principles,* which he otherwise praised. According to Smith, by using "Nature" instead of "God," and by taking such slight notice of biblical revelation, Lyell laid himself open to censure and suspicion. Smith says that he hopes his own work will provide the "explanations and corrections" which Lyell has failed to make in subsequent editions. See John Pye Smith, *On the Relation between the Holy Scriptures and Some Parts of Geological Science,* 5th ed. (London: Bohn, 1854), 208–9.

15. Buckland uses Lyell as a supporting reference for the length of geological time and for his denial of the transmutation of species, and he adopts Lyell's division of the tertiary strata. But Buckland also coopts Lyell in references to the catastrophic history of the earth. For example, in his discussion of fossiliferous strata, Buckland asserts that fossils are "documents which contain the evidences of revolutions and catastrophes, long antecedent to the creation of the human race," conveniently passing over Lyell's contrary explanation of slow and gradual processes. At another point, Buckland introduces the subject of volcanoes and earthquakes by referring to Lyell's comment that though the earth seems tranquil and solid enough to an Englishman, inhabitants of regions such as Italy would be of a very different opinion. Lyell's point is of course a localist one, that it is dangerous to extrapolate from one's immediate area, but Buckland uses it as a springboard for his catastrophist assertion about the formerly uninhabitable earth being made "the convenient and delightful habitation of man." (See 1:46–51, 98–105).

16. [Edward Forbes], "The Future of Geology," *Westminster Review* 58 (1852): 39. For information on Forbes, see Eric L. Mills, "A View of Edward Forbes, Naturalist," *Archives of Natural History* 11 (1984): 365–93. Forbes did fieldwork with Lyell in 1840–41, and his work was appropriated both by Lyell in the *Principles* and by Darwin in the *Origin.*

17. Ibid., 44.

18. The catalogue of Eliot and Lewes' library contains the 1851 edition of Lyell's *Manual* (the one reviewed by Forbes) and the tenth edition of the

*Principles* (the first to reflect Lyell's guarded acceptance of evolution and progressionism). See William Baker, *The Libraries of George Eliot and George Henry Lewes,* English Literary Studies Monograph Series 24 (Victoria, BC: ELS, 1981).

19. [G. H. Lewes], rev. of *Advanced Textbook of Geology,* by David Page, *The Leader* 7 (1856): 1192.

20. G. H. Lewes, prolegomena, *The History of Philosophy from Thales to Comte,* 3d ed., 2 vols. (London: Longmans, 1867), secs. 25–26.

21. G. H. Lewes, "Sea-Side Studies, Part III," *Blackwood's Edinburgh Magazine* 80 (1856): 473–74.

22. [G. H. Lewes], "Philosophy at Cambridge," rev. of *A Discourse on the Studies of the University of Cambridge,* by Adam Sedgwick, *The Leader* 1 (1850): 566-67.

23. "Lyell and Owen on Development," *The Leader* 2 (1851): 996.

24. G. H. Lewes, *Comte's Philosophy of the Sciences* (London: Bohn, 1853), 263.

25. Two entries from Somerville's book appear in the notebooks, and one of the entries is adjacent to the important material about floods copied from the *Annual Register.* See *George Eliot: A Writer's Notebook, 1854–1879,* ed. Joseph Wiesenfarth (Charlottesville: U of Virginia P, 1981), 36–38.

26. Mary Somerville, *Physical Geography,* 3d ed. (Philadelphia: Blanchard, 1853), 14, 24–25, 38, 56. This is the edition from which Eliot copied.

27. George Eliot, *The Mill on the Floss,* ed. Gordon S. Haight, Riverside Edition (1860; Boston: Houghton, 1961), bk. 1, ch. 7. All subsequent references are cited in the text.

28. I refer to the narrator as "her" because this opening chapter contains no gender identification of the narrator, and because we cannot read *The Mill* today without making associations among Eliot, Maggie, and the narrator.

29. W. J. Harvey, *The Art of George Eliot* (London: Chatto, 1963), 137.

30. Charles Lyell, *The Principles of Geology,* 3 vols. (London: Murray, 1830–33), 3:128. Subsequent references are cited parenthetically in the text.

31. Lyell, *Geological Evidences of the Antiquity of Man,* 405, 412.

32. Shuttleworth (*George Eliot and Nineteenth-Century Science,* 58–59) exposes the tenuousness of the narrator's argument. My only disagreement with Shuttleworth here is that I think she is too quick to put Eliot in the "progress" camp, with the result that passages such as this are seen as naive and inconsistent, rather than as part of an exploration of the question of progress in a variety of areas, including human lives, the evolution of organic beings and of human society, and the history of the earth.

33. Ibid., 58–62. For Eliot's criticism of progress, and an analysis of Darwinian views in her later novels, see K. M. Newton, "George Eliot, George Henry Lewes, and Darwinism," *Durham University Journal,* n.s. 35 (1974): 278–93.

34. Richard Owen, *On the Classification and Geographical Distribution of the Mammalia, to which is added "On the Gorilla" and "On the Extinction and Transmutation of Species"* (London: Parker, 1859), 102.

35. Ibid., 58.

36. Ibid., 58, 60.

37. Hardy, *Novels of George Eliot*, 57.

38. Hardy, *Particularities*, 63.

39. *Letters*, 3:317–18. Both Shuttleworth and Kerry McSweeney, "The Ending of *The Mill on the Floss*," *English Studies in Canada* 12 (1986): 55–68, see in the final chapter and the conclusion two different endings to the novel. I believe that we must look at these two "endings" as one unit.

40. We can even see this transformation in Eliot's manipulation of source material. We know that she researched locales and copied accounts of inundations from the *Annual Register* into her notebooks. Yet while Eliot drew many details for the flood from these passages, her imagined account is far less catastrophic than the journalistic accounts of these real floods.

41. During his adult life, Darwin was a voracious reader of novels, but he preferred those with happy endings; he read Eliot but, with the exception of *Silas Marner*, did not much care for her work. See Charles Darwin, *The Autobiography of Charles Darwin*, ed. Nora Barlow (1958; London: Pickering, 1989), 158; vol. 29 of *The Works of Charles Darwin*, ed. Paul H. Barrett and R. B. Freeman, 29 vols. (London: Pickering, 1986–89), and *The Life and Letters of Charles Darwin*, ed. Francis Darwin, 2 vols. (New York: Appleton, 1888), 1:102. For the argument that Darwinism's form is that of comic reversal, see A. Dwight Culler, "The Darwinian Revolution and Literary Form," *The Art of Victorian Prose*, ed. George Levine and William Madden (New York: Oxford UP, 1968, 224–46.

42. For Eliot's own dissatisfaction with her combination of these traditions, see *Letters*, 3:317–18. See also A. S. Byatt's introduction to the Penguin edition of the novel, and Ian Adam's "The Ambivalence of *The Mill on the Floss*," *George Eliot: A Centenary Tribute*, ed. Gordon S. Haight and Rosemary T. VanArsdel (Totowa, NJ: Barnes, 1982), 122–36.

43. Henry Adams, *The Education of Henry Adams* (1918; Boston: Houghton, 1961), 227.

44. This precise location of these earlier floods provides further evidence of Eliot's commitment to realistic, uniformitarian detail. The floods she read about while researching *The Mill* took place in the coastal regions of northeastern England in the late 1770s and were caused, like the flood in the novel, by days of heavy rains. Since the novel's flood occurs in 1839, and since the chronology is consistently and carefully marked, the great flood the old men remember from "sixty years ago" was in fact part of the same floods Eliot had read about.

45. Shuttleworth, *George Eliot and Nineteenth-Century Science*, 63.

46. [Lewes], rev. of *Advanced Textbook of Geology*, 1192.

47. This notebook material is available in K. K. Collins, "Questions of Method: Some Unpublished Late Essays," *NCF* 35 (1980): 390. Shuttleworth (*George Eliot and Nineteenth-Century Science*, 63) refers to this notebook entry as being from the same period as *The Mill* and uses it as an indication of Eliot's "departure from uniformitarian principles," but Collins believes the notebook

is probably from the *Middlemarch-Daniel Deronda* period, i.e., between 1869 and 1876.

## CHAPTER 5. RUSKIN'S "ANALYSIS OF NATURAL AND PICTORIAL FORMS"

1. *The Complete Works of John Ruskin,* Library Edition, ed. E. T. Cook and Alexander Wedderburn, 39 vols. (London: George Allen, 1903–12), 26:xxvi. All references to Ruskin's works will be cited parenthetically in the text.

2. John D. Rosenberg, *The Darkening Glass: A Portrait of Ruskin's Genius* (New York and London: Columbia UP, 1961), 11.

3. Patricia Ball, *The Science of Aspects: The Changing Role of Fact in the Work of Coleridge, Ruskin, and Hopkins* (London: Athlone, 1971), 65.

4. George P. Landow, *The Critical and Aesthetic Theories of John Ruskin* (Princeton: Princeton UP, 1971), 34.

5. Rosenberg, *Darkening Glass,* 10.

6. Ibid., 180.

7. Frederick Kirchoff, "A Science against Sciences: Ruskin's Floral Mythology," *Nature and the Victorian Imagination,* ed. U. C. Knoepflmacher and G. B. Tennyson (Berkeley: U of California P, 1977), 250–51.

8. Robert Hewison, *John Ruskin: The Argument of the Eye* (Princeton: Princeton UP, 1976), 20.

9. Quoted in ibid., 20. Like Rosenberg, Hewison sees Ruskin's "clinging" (21) to Saussure as a measure of his resistance to contemporary science.

10. Quoted in John Tyndall, *Essays on the Use and Limit of the Imagination in Science,* 2d ed. (London: Longmans, 1870), 2–3.

11. For a full account of the controversy, and Ruskin's role in it, see *Works,* 26:xxxiii–xli.

12. John Tyndall, *The Forms of Water in Clouds and Rivers, Ice and Glaciers* (New York: Appleton, 1872), sec. 95.

13. Elizabeth K. Helsinger, *Ruskin and the Art of the Beholder* (Cambridge: Harvard UP, 1982), 64.

14. In *The Poison Sky: Myth and Apocalypse in Ruskin* (Athens: Ohio UP, 1982), Raymond Fitch reads Ruskin's comments on Tyndall's lecture as full of irony and bitterness at what "must have seemed a crowning insolence of science" (543). While I agree with Fitch's view of the importance of *The Queen of the Air* for Ruskin's use of myth against mechanistic science, I see no basis for such a sweeping assumption about the passage's irony.

15. Jeffrey L. Spear, "'My darling Charles': Selections from the Ruskin-Norton Correspondence," *The Ruskin Polygon: Essays on the Imagination of John Ruskin,* ed. John Dixon Hunt and Faith M. Holland (Manchester: Manchester UP, 1982), 251.

16. John Tyndall, "On Chemical Rays, and the Light of the Sky," *Notices of the Proceedings at the Meetings of the Members of the Royal Institution of Great Britain* 5 (1866–69): 432.

17. Ibid., 440.
18. Ibid., 445.
19. Oliver J. Lodge, "Dust," *Nature* 31 (1884–85): 267.
20. Ibid., 265.
21. In Tyndall's experiments, the air introduced into the tube was first filtered and dried, and then passed through an amyl nitrate solution. Tyndall's purpose was to prove that "particles of infinitesimal size [in this case, the molecules of amyl nitrate vapor] without any colour of their own, and irrespective of those optical properties exhibited by the substance in a massive state, are competent to produce the colour of the sky" ("On Chemical Rays," 441). Ruskin interpreted this more narrowly to mean that the blue of the sky is caused by the air itself rather than by water vapor. Lodge, although basing his work on Tyndall's, wanted to show that in unfiltered air the "infinitesimal particles" that cause the sky to appear blue are dust particles, and that these dust particles are also the nuclei on which water vapor in the atmosphere condenses to form clouds, mist, and rain. Ruskin, apparently conflating these two claims, thus inferred that Lodge was ascribing the blue of the sky to water in the atmosphere.
22. See David B. Wilson, "A Physicist's Alternative to Materialism: The Religious Thought of George Gabriel Stokes," *Energy and Entropy: Science and Culture in Victorian Britain*, ed. Patrick Brantlinger (Bloomington: Indiana UP, 1989), 177–204.
23. George Gabriel Stokes, "On the Change of Refrangibility of Light," *Mathematical and Physical Papers*, 5 vols. (Cambridge: Cambridge UP, 1901), 3:270, 388.

## CHAPTER 6. "EUCLID HONOURABLY SHELVED": EDWIN ABBOTT'S *FLATLAND* AND THE METHODS OF NON-EUCLIDEAN GEOMETRY

1. *Science* 5 (1885): 184.
2. *Science* 5 (1885): 265.
3. R[obert] Tucker, "Flatland," *Nature* 31 (1884–85): 76.
4. J. J. Sylvester, "A Plea for the Mathematician," *Nature* 1 (1869): 237-38.
5. W. K. Clifford, "The Philosophy of the Pure Sciences," *Lectures and Essays*, ed. Leslie Stephen and Frederick Pollock, 3d ed., 2 vols. (London: Macmillan, 1901), 1:356.
6. Ibid., 1:387–88.
7. Articles in this dispute include Helmholtz, "The Axioms of Geometry," *The Academy* 1 (1870): 128–31; William Stanley Jevons, "Helmholtz on the Axioms of Geometry," *Nature* 4 (1871): 481–82; Helmholtz, "The Origin and Meaning of Geometric Axioms," *Mind* 1 (1876): 301–21; J. P. N. Land, "Kant's Space and Modern Mathematics," *Mind* 2 (1877): 38–46; Helmholtz, "The Origin and Meaning of Geometric Axioms, II," *Mind* 3 (1878): 212–25.
8. Helmholtz, "The Axioms of Geometry," 128.
9. For historical discussions of non-Euclidean geometry and its reception, see Jeremy Gray, "The Discovery of Non-Euclidean Geometry," *Studies in the*

*History of Mathematics*, ed. Esther R. Phillips, Studies in Mathematics 26 (n.p.: Mathematical Association of America, 1987), 37–60; Gray, *Ideas of Space* (Oxford: Clarendon, 1979); Marvin Jay Greenberg, *Euclidean and Non-Euclidean Geometries: Development and History*, 2d ed. (San Francisco: Freeman, 1980); Joan L. Richards, "The Evolution of Empiricism: Hermann von Helmholtz and the Foundations of Geometry," *British Journal for the Philosophy of Science* 28 (1977): 235–53; Richards, *Mathematical Visions: The Pursuit of Geometry in Victorian England* (San Diego: Academic P, 1988); Richards, "The Reception of a Mathematical Theory: Non-Euclidean Geometry in England, 1868-1883," *Natural Order: Historical Studies of Scientific Culture*, ed. Barry Barnes and Steven Shapin (Beverly Hills and London: Sage, 1979), 143–66; Richard J. Trudeau, *The Non-Euclidean Revolution* (Boston: Birkhäuser, 1987).

10. Riemann's lecture was not published until 1867 and not translated into English (by Clifford) until 1873. "On the Hypotheses Which Lie at the Bases of Geometry," *Nature* 8 (1873): 14–17, 36–37, rpt. in W. K. Clifford, *Mathematical Papers*, ed. Robert Tucker (1882; New York: Chelsea, 1968), 55–71.

11. Bertrand Russell, *An Essay on the Foundations of Geometry* (1897; New York: Dover, 1956), 1.

12. Helmholtz, "Origin and Meaning," 302, 319.

13. Hastings Berkeley, *Mysticism in Modern Mathematics* (London: Frowde, 1910), 219.

14. Helmholtz, "Axioms of Geometry," 128.

15. Helmholtz, "Origin and Meaning," 305.

16. Ibid., 316.

17. Ibid., 314.

18. O[laus] Henrici, "The Axioms of Geometry," *Nature* 29 (1884): 453–54; Henri Poincaré, "Non-Euclidean Geometry," *Nature* 45 (1892): 405; Simon Newcomb, "Modern Mathematical Thought," *Nature* 49 (1894): 328; W. G., "Euclid, Newton, Einstein," letter, *Nature* 104 (1920): 627-30; Dionys Burger, *Sphereland*, trans. Cornelie J. Rheinboldt (New York: Crowell, 1965); Cornelius Lanczos, *Space through the Ages* (London: Academic P, 1970); Rudolf v. B. Rucker, *Geometry, Relativity and the Fourth Dimension* (New York: Dover, 1977).

19. G. H. Lewes, "Imaginary Geometry and the Truth of Axioms," *The Foundations of a Creed*, 2 vols. (London: Trübner, 1874-75), vol. 2, app. A.

20. Richards, "Reception," 146-60; "Evolution," 247–52; *Mathematical Visions*, 3–11, 26–39, 56–57, 85–113.

21. Arthur Cayley, "Presidential Address to the British Association, September 1883," *The Collected Mathematical Papers of Arthur Cayley*, 13 vols. (Cambridge: Cambridge UP, 1896), 11:430. Subsequent references are cited in the text.

22. Russell, *Essay on the Foundation of Geometry*, 101.

23. Ibid., 105. Russell was, however, familiar with popular works about the fourth dimension and the implications of non-Euclidean geometry, reviewing, for example, C. H. Hinton's *The Fourth Dimension* in *Mind*, n.s. 13 (1904): 573–74.

24. Rosemary Jann, "Abbott's *Flatland:* Scientific Imagination and 'Natural Christianity,'" *Energy and Entropy: Science and Culture in Victorian Britain*, ed.

Patrick Brantlinger (Bloomington: Indiana UP, 1989), 289–306; Elliot Gilbert, "*Flatland* and the Quest for the New," *ELT* 34 (1991): 391–404; Thomas Banchoff, "From *Flatland* to Hypergraphics," *Interdisciplinary Science Reviews* 15 (1990): 364–72, and his introduction to the Princeton Science Library's edition of *Flatland* (Princeton: Princeton UP, 1991). Jann argues that "*Flatland* alludes to contemporary debate over the role of hypotheses in scientific discovery and the relationship between material proof and religious faith" (289), that Abbott is "trying to keep the new territories of science hospitable to humanity's higher needs by forging an alliance between the theoretical and the spiritual" (290). I want to expand on the connection to the debate over scientific method by looking at Abbott's role, and the role of non-Euclidean geometry, in that debate more closely. Jann's focus is more on Abbott's theology and the cultural relations of science and religion. Subsequent references to Jann are cited in the text.

25. For biographical information on Abbott, see his obituary in the London *Times* (13 October 1926); A. E. Douglas-Smith, *The City of London School*, 2d ed. (Oxford: Blackwell, 1965), and the works by Banchoff, who is writing a biography.

26. James Spedding, "The Latest Theory about Bacon," *Contemporary Review* 27 (1875–76), 653; notices of Spedding's article, with rebuttal letters from Abbott, appear in *Spectator* 49 (1876): 472, 498–99 and *Academy* 9 (1876): 333, 360; Abbott, "The Latest Theory about Bacon," *Contemporary Review* 28 (1876): 141–67; Spedding, "Lord Macaulay's Essay on Bacon Examined," *Contemporary Review* 28 (1876): 169–90, 365–92, 562–86; Abbott, *Bacon and Essex* (London: Seeley, 1877); [Agnes M. Clerke], "Spedding's *Life of Bacon*," *Edinburgh Review* 150 (1879): 395–436; Spedding, "Dr. Abbott and Queen Elizabeth," *Nineteenth Century* 7 (1880): 107–27.

27. Edwin A. Abbott, introduction, *Bacon's Essays*, 7th ed., 2 vols. (London: Longmans, 1886), 1:xxiv. Subsequent references are noted parenthetically in the text.

28. Edwin A. Abbott, *Francis Bacon: An Account of His Life and Works* (London: Macmillan, 1885), 344. Subsequent references will be noted parenthetically in the text.

29. Abbott, *The Kernel and the Husk* (Boston: Roberts, 1887), 17. Further references in the text.

30. Samuel Taylor Coleridge, *Logic*, ed. J. R. de J. Jackson (Princeton: Princeton UP, 1981), 136–37.

31. Douglas-Smith, *City of London School*, 94, 125.

32. Sylvester, "Plea for the Mathematician," 261.

33. See Richards, *Mathematical Visions*, 161–98.

34. Russell, *Essay on the Foundations of Geometry*, 105.

35. A. T. Schofield, *Another World, or The Fourth Dimension* (1888; London: Sonnenschein, 1890), 3–4.

36. Hinton wrote on the fourth dimension as early as 1880, but published his own version of *Flatland* in *An Episode of Flatland: or, How a Plane Folk Dis-*

*covered the Third Dimension* in 1907. His nonfictional works include *A New Era of Thought* (1888), *The Fourth Dimension* (1904), and *A Language of Space* (1906).

37. Abbott, *Apologia* (London: Black, 1907), 11–12.

38. Ibid., 63.

39. Ibid., 14n. Edwin A. Abbott, *Flatland* (1963; New York: Barnes, 1972), xv. Subsequent references will be cited parenthetically in the text.

40. In *English Lessons for English People* (Boston: Roberts, 1893), Abbott and co-author J. R. Seeley state that "The Argument from Analogy . . . so far as it is an argument at all, comes under the head of Induction. Otherwise it is not an argument, but a metaphorical illustration of an argument" (sec. 190). For insight on Abbott's attitude to the argument from analogy in *Flatland*, I am indebted to conversations with Lawrence Berkove and Gerald Baker.

41. There is bite in this "Wentbridge" joke on at least two levels. The study of Euclid had long been part of the gentlemanly education at the universities, whereas Abbott, though a Cambridge man himself, was associated with the City of London School's more middle-class, mercantile, and practical educational mission. More specifically, Oxbridge resisted reform efforts like the AIGT's (see Richards, *Mathematical Visions*, 174). Abbott's satire is thus directed at institutions whose teaching of geometry and methodological understanding of mathematics is, from Abbott's perspective, out of date and possibly headed in the wrong direction—and hence the grammarian Abbott's turning of Cambridge into Wentbridge?

42. Of his effort to compose "a treatise on the mysteries of Three Dimensions" (103) for his fellow Flatlanders, A Square recalls his representational difficulties: "I spoke . . . of a Thoughtland whence, in theory, a Figure could look down upon Flatland and see simultaneously the insides of all things. . . . But in writing this book I found myself sadly hampered by the impossibility of drawing such diagrams as were necessary for my purpose; . . . so that, when I finished my treatise (which I entitled, 'Through Flatland to Thoughtland' [compare the title of Abbott's 1877 *Through Nature to Christ*]) I could not feel certain that many would understand my meaning" (104).

43. For a discussion of the development of the necessary truth of mathematics even among British empiricists, see Joan L. Richards, "God, Truth, and Mathematics in Nineteenth Century England," *The Invention of Physical Science: Intersections of Mathematics, Theology, and Natural Philosophy since the Seventeenth Century*, ed. Mary Jo Nye, Joan L. Richards, and Roger H. Stuewer (Dordrecht: Kluwer, 1992), 51–78.

44. Samuel Taylor Coleridge, *Biographia Literaria*, ed. James Engell and W. Jackson Bate, 2 vols. (Princeton: Princeton UP, 1983), 1:250.

CHAPTER 7. "THE METHODS HAVE BEEN OF INTEREST": SHERLOCK HOLMES, SCIENTIFIC DETECTIVE

1. Arthur Conan Doyle, *Through the Magic Door* (London: Smith, 1907), 253.

2. Arthur Conan Doyle, *Memories and Adventures* (1924; London: Greenhill, 1988), 26, 74–75.

3. Ibid., 108.

4. Willard Huntington Wright, "The Great Detective Stories," *The Art of the Mystery Story,* ed. Howard Haycraft (New York: Simon, 1946), 59. Wright's essay first appeared in 1927.

5. Edmund Wilson, "'Mr. Holmes, They Were the Footprints of a Gigantic Hound,'" *The Baker Street Reader,* ed. Philip A. Shreffler (Westport, CT: Greenwood, 1984), 37.

6. John G. Cawelti, *Adventure, Mystery, Romance* (Chicago: U of Chicago P, 1976), 11, 93. Similar positions are adopted by Pierre Nordon, *Conan Doyle: A Biography* (1964; trans., New York: Holt, 1967), 217–19, and J. K. Van Dover, "The Lens and the Violin: Sherlock Holmes and the Rescue of Science," *Clues* 9.2 (1988): 37–51. For a less polarized position, see Joseph Agassi, "The Detective Novel and Scientific Method," *Poetics Today* 3 (1982): 99–108.

7. "The Red-Headed League." Subsequent references will appear in the text using the following standard abbreviations for the Holmes stories:

| | | | |
|---|---|---|---|
| ABBE | Abbey Grange | REDH | Red-Headed League |
| BLAN | Blanched Soldier | REIG | Reigate Puzzle |
| BLUE | Blue Carbuncle | SCAN | Scandal in Bohemia |
| BOSC | Boscombe Valley Mystery | SIGN | Sign of Four |
| COPP | Copper Beeches | SILV | Silver Blaze |
| CROO | Crooked Man | SPEC | Speckled Band |
| DANC | Dancing Men | STUD | Study in Scarlet |
| FIVE | Five Orange Pips | SUSS | Sussex Vampire |
| HOUN | Hound of the Baskervilles | 3GAR | Three Garridebs |
| IDEN | A Case of Identity | TWIS | Man with Twisted Lip |
| NAVA | Naval Treaty | VALL | Valley of Fear |
| NORW | Norwood Builder | | |

Part and chapter numbers will be given for the four novellas.

8. Nordon, *Conan Doyle,* 244.

9. Ibid. Ian Ousby, *Bloodhounds of Heaven* (Cambridge: Harvard UP, 1976), 154.

10. Régis Messac, *Le «Detective Novel» et l'influence de la pensée scientifique* (1929; Geneva: Slatkine, 1975), 13, 602–17.

11. Ousby, *Bloodhounds of Heaven,* 153; Nordon, *Conan Doyle,* 247; James Kissane, and John M. Kissane, "Sherlock Holmes and the Ritual of Reason," *NCF* 17 (1963): 356–57; J. L. Hitchings, "Sherlock Holmes the Logician," *Baker Street Journal* 1 (1946): 113–17.

12. See Thomas A. Sebeok and Jean Umiker-Sebeok, *"You Know My Method": A Juxtaposition of Charles S. Pierce and Sherlock Holmes* (Bloomington, IN: Gaslight, 1980), and Umberto Eco and Thomas Sebeok, eds., *The Sign of Three: Dupin, Holmes, Peirce* (Bloomington: U of Indiana P, 1983). Peirce and Doyle were contemporaries, and Peirce's writings on method are a major American contribution to the nineteenth-century movement away from Baconian induction. We owe to these semiotic critics a thorough presentation of Holmes's methodological statements, yet the linkage of Holmes and Peirce often comes across as forced: without evidence for Doyle's familiarity with Peirce, ignoring Doyle's documented interest in other methodologists is hard to justify.

13. Gian Paolo Caprettini, "Peirce, Holmes, Popper," *Sign of Three*, 140.

14. In his discussion of Holmes's character, Ian Ousby (*Bloodhounds of Heaven*, 151–59) focuses on the changes that appear over time, arguing that Doyle's conception of Holmes alters significantly, especially after the earliest novellas and stories. In general terms, Ousby makes a convincing case, and I would suggest that a similar movement can be discerned in the presentation of Holmes's methods, with growing emphasis placed on intuition, speculation, etc. But I emphasize that the juxtaposition of these seemingly disparate features of both character and method occur throughout the Holmes canon; they cannot be separated neatly into different phases, nor do they follow a linear developmental pattern. Indeed, it is striking that the comments about Holmes that Doyle makes in *Memories and Adventures*, written near the end of the cycle, exactly reproduce comments made by Watson in the earliest stories. In light of this, it seems to me significant that one area in which a fairly radical break is evident is in the connection between the scientific and the inhuman, the unemotional.

15. Messac (*Le "Detective Novel,"* 613–15) was the first to make this point. For him, these episodes are "puerile games" that have nothing to do with science; the model provided by Bell is part of the artistic rather than scientific presentation of Holmes.

16. Doyle, *Memories*, 31.

17. T. H. Huxley, "On the Method of Zadig" (1880), *Collected Essays*, 9 vols. (New York: Greenwood, 1968), 4:1–23. Subsequent references to Huxley's works are cited in the text as *Collected Essays*. I am not the first to note the affinities between Holmes and Huxley or Holmes and Zadig; my endeavor here is to emphasize the deep interconnections and detailed correspondences. See Messac, *Le "Detective Novel"*, 33–38; Ousby, *Bloodhounds of Heaven*, 154; Jacques Barzun, "Detection and the Literary Art, "*The Delights of Detection*, ed. Jacques Barzun (New York: Criterion, 1961), 12; Eco, "Horns, Hooves, Insteps: Some Hypotheses on Three Types of Abduction," *Sign of Three*, 198–220.

18. *Collected Essays*, 3: 45–46 (original emphasis). Huxley employs the example again in "The Method by Which the Causes of the Present and Past Conditions of Organic Nature Are to Be Discovered," one of his "Six Lectures to Working Men" in 1862: "I will suppose that one of you, on coming down

in the morning to the parlour of your house, finds that a tea-pot and some spoons which had been left in the room on the previous evening are gone,—the window is open, and you observe the mark of a dirty hand on the window-frame, and perhaps in addition to that, you notice the impress of a hob-nailed shoe on the gravel outside. All these phenomena have struck your attention instantly, and before two seconds have passed you say, 'Oh, somebody has broken open the window, entered the room, and run off with the spoons and the tea-pot!' That speech is out of your mouth in a moment. And you will probably add, 'I know there has; I am quite sure of it!' You mean to say exactly what you know; but in reality you are giving expression to what is, in all particulars, an hypothesis. You do not *know* it at all; it is nothing but an hypothesis rapidly framed in your own mind. And it is an hypothesis founded on a long train of inductions and deductions" (2:368–69).

The insistence that science is organized common sense is echoed almost verbatim by Holmes: "my simple art . . . is but systematized common sense" (BLAN). It is crucial, I think, to see this view as one of the lingering influences of the Baconian program, in which genius is not necessary for scientific achievement. Part of the movement away from this program during the course of the century is due to the increasing awareness that science was often anything but commonsensical. Ed Block, Jr.'s "T. H. Huxley's Rhetoric and the Popularization of Victorian Scientific Ideas, 1854–1874," *Energy and Entropy: Science and Culture in Victorian Britain*, ed. Patrick Brantlinger (Bloomington: Indiana UP, 1989), 205–28, discusses the tensions within Huxley's work between this commonsense view and the very uncommonsensical ideas he was trying to popularize. We have seen that a similar tension infuses Holmes's methodology.

19. Doyle, *Through the Magic Door*, 250.

20. John Tyndall, "Scientific Materialism," *Fragments of Science*, 6th ed., 2 vols., (New York: Burt, n.d.), 1:400, 405, 407. Subsequent references are cited in the text.

21. Oscar Wilde, *The Picture of Dorian Gray*, ed. Peter Ackroyd (Harmondsworth: Penguin, 1985), 21. Doyle met Wilde at a dinner given by publisher J. M. Stoddart in 1889; it was at this dinner that Doyle agreed to submit "The Sign of Four," Wilde *The Picture of Dorian Gray*, to Stoddart's *Lippincott's Monthly Magazine*. See Doyle, *Memories*, 78–79.

22. Oscar Wilde, *The Complete Writings of Oscar Wilde*, 9 vols. (New York: Nottingham Society, 1909), 7:44, 191, 205. Subsequent references are noted in the text.

23. Arthur Conan Doyle, preface to *The Case Book of Sherlock Holmes, The Complete Sherlock Holmes*, 2 vols. (Garden City, NY: Doubleday, n.d.) 2:984.

24. Agassi, "Detective Novel and Scientific Method," 105.

25. Ibid., 108.

26. Another term might be Tyndall's "supplementary." John Hannay has recently used "complementarity" in precisely the opposite sense from that which I propose, to affirm his "belief in the incompatibility . . . between science

and literature." See John Hannay, editor's comment, *The Literary Uses of the Rhetoric of Science*, spec. issue of *Studies in the Literary Imagination* 22.1 (1989): 1.

27. Caprettini, "Peirce, Holmes, Popper," 144 (original emphasis).

28. Albert Einstein and Leopold Infeld, *The Evolution of Physics* (New York: Simon, 1938), 4–5.

# Index


Abbott, Edwin: and *Flatland*, 9, 180–200 *passim*, 201–8, 264*n24*; on non-Euclidean geometry, 183, 184, 186, 188, 197–201, 202, 206–8, 209, 210, 264*n24*; assessment by, of Baconian induction, 191–94, 197, 209; and Tyndall's "scientific use of the imagination," 194, 195, 200; importance of Imagination in his epistemology, 194–98 *passim*; Coleridge's influence on, 195–97, 210. *The Kernel and the Husk*, 194, 195, 198, 199

Abrams, M. H., 49, 53, 250*n23*

Adams, Henry, 146, 147, 149, 151, 244*n48*

Agassi, Joseph, 235

Arac, Jonathan, 251*n34*

Bacon, Francis: his induction assessed, 4, 11–44 *passim*, 242*n21*, 243*n35*, 247*n103*; as Moses figure, 11–12, 37, 40–42, 193, 247*n108*; Coleridge's view of, 16, 72–77, 251*nn32,34*; Wordsworth's view of, 52, 53; Davy's view of, 81–84, 87; Ruskin's view of, 163, 164, 179; Abbott's view of, 191–94, 197, 198. *Novum Organum*, 4, 12, 13, 18, 26, 30, 32, 38–43 *passim*, 72, 75, 76, 193, 194. *See also* Baconian induction, assessments of

Baconian induction: debate over, place in nineteenth-century culture, 5, 13, 15, 43–44; and Newton, 12, 17, 18, 28–33, 42–43; in eighteenth century, 12, 18, 245*n70*; naive version of, 13–19, 20–44 *passim*; example of heat in, 14, 29, 245*n73*; defined, 14–16; facts in, 16–17, 19–23, 38–39; and deduction, 17–18, 25, 31–32, 39; observation in, 20–23; objectivity in, 23–24; elimination in, 24–27; enumeration in, 24–27; imagination or genius in, 25–28, 33–37; and common sense, 27–28; hypotheses in, 28–33, 35; and geology, 97–99, 122, 135; and Sherlock Holmes, 215–20 *passim*, 230, 231, 232, 236. *See also* Naive Baconianism

—assessments of: by Abbot, 192–94, 197, 209; by Brewster, 18, 25, 29; by Brougham, 28; by Coleridge, 16, 21, 63–64, 72–77, 251*n32*; by Comte, 18, 31; by Davy, 12, 14–15, 83–84; by De Morgan, 18, 24; by Ellis, 4, 19, 22, 27, 42; by *Encyclopaedia Britannica*, 12, 42; by Fischer, 18; by Fowler, 13, 29, 30–31, 42; by Hallam, 13, 41; by Herschel, 23, 25, 245*n5*; by Jevons, 13, 18, 19, 24, 26, 32, 243*n35*; 245*n73*; by Lewes, 17, 21, 30, 42; by Liebig, 22, 42; by Macaulay, 17, 27, 41–42, 242*n21*; by Merz, 42–43; by Mill, 22, 26, 30, 31, 32; by Nichol, 25, 42; by Pearson, 23; by Playfair, 12; by Powell, 40–41, 245*n73*; by Robertson, 13; by Ruskin, 163, 164; by Whewell, 21–31 *passim*, 37–38, 40; by Wordsworth, 52, 53

Ball, Patricia, 158

Bell, Dr. Joseph, 211, 224, 225

Black, Joel, 46–47

Brewster, David, 18, 20, 25, 26, 29

Bridgewater Treatises, 124, 125, 126, 139, 142, 149, 258*n13*

British Association for the Advancement of Science, 25, 33, 35, 44, 154, 168, 174, 181, 188, 230

Brougham, Henry, 14, 25, 28, 88, 89
Buch, Leopold von, 113, 120
Buckland, William, 9, 96, 124, 125, 126, 128, 130, 142, 143, 148, 149, 150, 152, 258*nn13, 15*

Cannon, Susan F., 14, 112, 245*n57*
Caprettini, Gian Paolo, 236, 237
Carpenter, W. B., 35
Carroll, Lewis, 180
Cawelti, John, 212
Cayley, Arthur, 188, 189, 190, 198, 200, 208, 209, 210
Chambers, Robert, 129
Clifford, W. K., 182, 184, 187, 189, 197, 208, 209, 210, 263*n10*
Coleridge, Samuel Taylor: and Baconian induction, 16, 17, 21, 52, 58, 242*n15*, 251*n34*, 252*n40*; and Davy, 47, 79, 81, 82, 84, 85, 86, 89, 90, 250*n18*; and geology, 99, 100; and Abbott, 195–96, 198, 206, 210. Works: *Biographia Literaria*, 69, 71, 76, 251*n34*; *The Friend*, 8, 62, 67, 68, 72, 75, 76, 89, 251*n26*, 252*n40*; *Logic* 75, 252*n36*. *See also* Romanticism
—"Essay on Method": hostility to classification in, 63, 64, 65, 66; mind in, 63, 65, 70; method of fine arts in, 68; Law and Theory distinguished in, 66, 67, 68, 251*n31*; hypotheses in, 66–67; Imagination in, 69; Reason and Understanding distinguished in, 69, 70; relationship of poetry and science in, 71; Bacon and Plato reconciled in, 72–77, 252*n40*; Davy in, 76–77, 85–87, 89
Comte, Auguste, 17, 18, 21, 22, 31, 32, 35, 129, 130, 139
Conan Doyle, Arthur. *See* Doyle, Arthur Conan
Cowley, Abraham, 11, 17, 37, 40, 41, 42, 247*n108*
Cuvier, Georges, 224–31 *passim*, 237

Darwin, Charles: methods criticized, 3–5, 43–44; and Lyell, 9, 93, 94–95, 96, 122–24, 138; and *The Mill on the Floss*, 9, 93, 121, 122, 124, 132–50 *passim*, 260*n41*; and Romanticism, 93, 109–12, 256*nn39, 42*; Eliot's view of, 122, 129, 130, 139, 148; Ruskin's criticism of, 153, 170. Works: *Autobiography*, 94, 112; *Diary*, 106, 107, 108, 110, 118, 133, 255*n35*; *Origin of Species*, 94, 120, 121, 122, 123, 125, 138
—*Journal of Researches* (*Voyage of the Beagle*): uniformitarian geology in, 9, 93, 94–95, 103–20, 255*n35*; and Humboldt, 112–20; narrative structure of, 116–20. *See also* Geology
Davy, Humphry: on relationship of poetry and science, 8, 47, 49, 58, 77–82, 90, 250*n18*; and Wordsworth and Coleridge, 8, 47, 58, 62, 76, 89, 90, 250*n18*; on Baconian induction, 12, 14, 17, 18; on scientific method, 25, 83, 84, 85, 87, 88. Works: *Consolations in Travel*, 78, 80; *Elements of Chemical Philosophy*, 14, 83; "Introductory Discourse," 58, 82; "Lecture on Electro-Chemical Science," 84; "On the Safety Lamp," 87–88. *See also* Romanticism
De Beaumont, Elie, 98, 115
De Morgan, Augustus, 18, 24, 33, 43
De Quincey, Thomas: on literature of knowledge v. literature of power, 6–7, 50, 51, 52, 53, 71, 81; attitude of, to science, 7, 50, 51, 52, 53, 70, 71, 76, 81
Descartes, René, 29, 43
Dickens, Charles, 135
Dodd, George, 45, 90
Dowden, Edward, 6–8
Doyle, Arthur Conan, 10, 211, 212, 213, 214, 223–37 *passim*

Eco, Umberto, 214
Eichner, Hans, 46
Einstein, Albert, 90, 188, 237
Eliot, George: on Lyell, 122, 123, 125–30, 150; on Darwin, 122, 129, 130, 139; on natural theology, 124–128, 130, 139–41; on geology, 124–130
—*The Mill on the Floss*: uniformitarian geology in, 9, 93, 120, 121, 122, 124, 130–51; and Darwin, 9, 93, 121, 122,

124, 132–50 *passim*, 260*n41*; natural theology in, 9, 121, 140–41, 142–43, 148; ending of, 9, 121, 141, 143–45, 260*n39*; floods in, 9, 121–22, 132–33, 135, 141–45, 147–49, 150, 260*n40*; and Romanticism, 93, 100, 122, 133–34, 135, 138, 144–45
Eliot, T. S., 161
Ellis, Robert Leslie: Bacon's *Works*, editor of, 4, 19; assessments by, of Baconian induction, 4, 19, 20, 22, 27, 42, 216; and Abbott, 192, 194
*Encyclopaedia Britannica*, 12, 33, 35, 65
Euclid, 52, 180–86 *passim*, 197, 198, 200, 208, 209, 210, 232, 265*n41*. *See also* Geometry; Non-Euclidean geometry

Fischer, Kuno, 18
Forbes, Edward, 126, 127
Forbes, J. D., 169, 171
Foucault, Michel, 59, 60, 85
Fowler, Thomas, 13, 29, 30, 31, 42

Geology: and Lyell, 9, 92–103, 104–16 *passim*, 119–20, 121–31 *passim*, 134, 135, 138, 144, 146–50; and Darwin, 9, 93, 94–95, 103–12, 113–24 *passim*, 129–30; and Eliot, 9, 121–51; and Ruskin, 152, 164–71 *passim*
—catastrophist: and Lyell, 9, 95, 100, 104–8 *passim*, 123, 125; and realism, 9, 100, 106, 107, 108, 120; and Darwin, 9, 104–8, 111, 120; and imagination, 9, 104–8, 111, 120; and Romanticism, 9, 111; Eliot's view of, 9, 121, 125–26, 130, 141–43; and *The Mill on the Floss*, 9, 121, 133, 135, 141–50 *passim*; method of, 95, 100, 123, 254*n15*; and natural theology, 95–96, 121, 123, 125–26, 130, 141–45 *passim*; and history, 105, 107, 133
—uniformitarian: and Lyell, 9, 92–93, 95–103, 104–21 *passim*, 122–24, 254*n15*; method of, 9, 92–93, 95–103, 122–23, 254*n15*; and induction, 9, 93, 97, 98, 99, 103, 122; and imagination, 9, 93, 97, 98, 103–12, 115, 119, 135; and realism, 9, 93, 97, 99, 112, 135; and Romanticism, 9, 93, 100, 109–12, 122, 134–35, 145; and *The Mill on the Floss*, 9, 93, 121, 122, 131–51; and Darwin, 9, 94–95, 96, 103–20, 122–24; and history, 92–93, 96–97, 99, 100–103, 105–20 *passim*, 131, 133–34; Whewell's view of, 95, 100; Coleridge's view of, 99–100; Eliot's view of, 125–30
Geometry, 9, 52, 70, 180–91, 197–210. *See also* Non-Euclidean geometry
Gilbert, Davies, 78
Goethe, Johann Wolfgang von, 18, 36, 46, 252*n40*
Gould, Stephen Jay, 95, 96, 98, 119

Haight, Gordon, 124
Hallam, Henry, 13, 18, 41
Hamilton, W. R. (Irish mathematician), 45, 61, 62
Harcourt, Vernon, 124
Hardy, Barbara, 121, 141
Harvey, W. J., 135
Helmholtz, Hermann von, 181–90 *passim*, 202, 207, 208, 209, 210
Helsinger, Elizabeth, 171, 172
Herschel, John: assessments by, of Baconian induction, 18, 23, 25, 27, 43, 245*n57*; and hypotheses, 29, 33, 34; on Mill-Whewell debate, 31; view by, of Lyell's methods, 92–93, 99–100. *Preliminary Discourse on the Study of Natural Philosophy*, 23, 31, 33, 92, 99, 179, 245*n57*
Hewison, Robert, 167
Hinton, C. H., 199, 209, 263*n23*
Holmes, Sherlock: personality of, 6, 211, 212, 213, 222, 223, 267*n14*; and Romanticism, 10, 212, 213, 232, 234, 235
—methods of: scientific, 10, 211, 212, 213–15; intuition in, 10, 212–13, 221–22, 230; artistic, 10, 212–13, 232–35; hypotheses in, 10, 217, 219, 220, 221; deduction in, 10, 218, 225, 226; as common sense, 28, 228, 229, 268*n18*; and Joe Bell, 211, 224; and Baconian induction, 215–16, 230, 232, 236; theory in, 217, 218, 219, 220; imagination in, 217–26 *passim*, 230, 232, 237; and chemistry, 225; and Cuvier, 225–30; and Huxley, 227–30; and Tyndall's "scientific use of the imagination," 230–32

Humboldt, Alexander von, 112–20
Hunt, Leigh, 49, 50, 51, 70, 71, 76
Hutton, James, 97, 98, 127
Huxley, Thomas Henry: rejection by, of naive Baconianism, 4, 17, 43, 44, 247*n103*; Darwin's method, defense of, 4, 44; and Sherlock Holmes, 10, 229–30; science as common sense, 28, 229, 267*n18*; materialism of, criticized, 153, 195; criticism by, of mathematics, 181, 209; method of, 227–29
Hyman, Stanley Edgar, 109
Hypotheses: and Sherlock Holmes, 10, 217–21 *passim*; and Baconian induction, 12, 28–37 *passim*; and Newton, 14, 17, 28–33, 176; Whewell on, 14, 29, 31, 34; Herschel on, 29, 33, 34; Tyndall on, 32–33, 168, 231–32; Coleridge on, 66–67; Davy on, 83–84, 86–87, 88

Imagination: Tyndall's "scientific use of," 8, 9, 34–35, 153, 154, 166–79 *passim*, 183, 194, 195, 200, 230, 231, 242*n12*, 246*n93*; and geology, 9, 93, 97, 98, 103–12, 115, 119, 120, 135; and Ruskin, 9, 153, 156, 160–79 *passim*; Whewell on, 25, 26, 27, 29, 33, 34; in Baconian induction, 25–28, 33–37; and Wordsworth, 57–58; and Coleridge, 68–69, 89; and Davy, 77–82, 88–89; Darwin's, 103–12; and non-Euclidean geometry, 181, 186–91, 208–10; Abbott on, 194–98 *passim*; of Sherlock Holmes, 217–26 *passim*, 230, 232, 237
Induction. *See* Baconian induction

James, Henry, 121, 141
Jann, Rosemary, 191, 194, 195, 204, 264*n24*
Jevons, W. Stanley, 13, 14, 18, 19, 20, 23, 26, 31, 32, 33, 187, 243*n35*, 247*n103*

Kant, Immanuel: and Coleridge, 69, 70, 72, 75, 76; Kantianism, 69, 70, 76, 179, 182, 184, 187, 209, 210; and non-Euclidean geometry, 182–88 *passim*, 209–10
Keats, John, 46, 56, 162, 163
Kirchoff, Frederick, 167

Lamarck, Jean-Baptiste de, 98, 122, 127
Landow, George, 162
Laplace, Pierre Simon, 51, 229, 230
Levine, George, 6
Lewes, George Henry: and Darwin, 4, 44, 122, 139; on scientific method, 17, 20–21, 29–30, 35–36; rejection by, of naive Baconianism, 17, 29–30, 42; and uniformitarianism, 128–29, 140, 149; and non-Euclidean geometry, 187–88, 189, 190. Works: *Comte's Philosophy of the Sciences*, 129; *History of Philosophy*, 128; *Life of Goethe*, 35–36; *Sea-Side Studies*, 128, 140
Liebig, Justus von, 22, 35, 42
Lodge, Oliver, 174, 175, 176, 178, 262*n21*
Lorrain, Claude, 153, 159, 160, 165, 172
Lovejoy, Arthur, 69
Lyell, Charles: and Darwin, 9, 93, 94–95, 96, 122–24, 138; criticism by, of natural theology, 95–96, 121–26 *passim*, 140–41, 148–49, 258*nn14,15*; Eliot's view of, 122, 123, 125–30, 150; and *The Mill on the Floss*, 122, 131–51 *passim*; and Lewes, 128–29. Works: *Antiquity of Man*, 124; *Elements of Geology*, 101; *Manual of Elementary Geology*, 126
—*Principles of Geology*: method of, 9, 92–93, 95–97, 98–103, 122–24; Darwin's *Journal of Researches* influenced by, 9, 103–8, 113, 114; and *The Mill on the Floss*, 9, 135, 138, 146; Darwin's reading of, 94–95; Eliot's view of, 124–27

Macaulay, Thomas Babington, 14, 17, 18, 27, 41, 70, 192
Materialism: and Tyndall, 9, 34–35, 195, 230–31, 235; Ruskin's criticism of, 9, 153, 154, 156, 162–63, 169–79 *passim*; Wordsworth's criticism of, 56–57; Coleridge's criticism of, 63–66, 69, 74–75, 76, 86; Huxley's, criticized, 153, 195;

Abbott's criticism of, 195, 196; and Doyle, 230, 235
Merz, John Theodore, 24, 42, 43
Messac, Régis, 214
Mill, John Stuart: on Darwin's method, 4, 44; and Baconian induction, 15, 17, 18, 22, 26, 30, 31, 32, 43; debate with Whewell, 31; and Ruskin, 179; and geometry, 184, 188, 189. *System of Logic*, 4, 31, 85, 179, 189, 229
Montagu, Basil, 17, 72, 252*n40*

Naive Baconianism: Huxley and, 4, 17, 43, 44, 247*n103*; assessments of, 8, 13–19, 20–44 *passim*; Romantic rejection of, 8, 46; Wordsworth and, 49, 52, 54, 56, 57, 61; Coleridge and, 63–64, 67, 70, 73–77 *passim*; Davy and, 77, 82–84, 87, 88, 89; Ruskin and, 156–64 *passim*, 167–68, 178–79, 247*n103*; Abbott and, 192–94, 197, 209; Sherlock Holmes and, 215–20 *passim*; Tyndall and, 231, 247*n103*. *See also* Baconian induction
Natural theology, 9, 86, 95, 96, 121–30 *passim*, 139–49 *passim*, 203, 204, 258*nn14, 15*
*Nature*, 45, 57, 90, 174, 181, 182, 186
Newton, Isaac: and Bacon, 8, 12, 14, 17, 18, 20, 28, 30, 32, 33, 37, 42, 43, 46, 83, 84; and hypotheses, 14, 17, 28–33, 176; Romantics' views of, 46, 51, 52, 76, 81, 82, 83, 84, 87, 91; Ruskin's view of, 163, 176. *Principia*, 32, 42, 51, 52, 53, 81
Nichol, John, 25, 42
Niebuhr, Barthold, 102, 103, 106, 133, 147, 255*n32*
Non-Euclidean geometry: and *Flatland*, 9, 180, 201–08; popularizations of, 180, 181, 184–86, 191; imagination in, 181, 186–91, 208–10; and Kantianism, 182, 184, 186, 189–90, 209–10; history of, 183–84; and spiritualism, 187, 199–201; and Abbott, 197–201, 208–10
Nordon, Pierre, 213, 214

Ousby, Ian, 214, 267*n14*
Owen, Richard, 9, 123, 129, 139, 140, 141
Owen, W. J. B., 59

Page, David, 128, 149
Paradis, James, 109, 111
Pearson, Karl, 23, 36, 37, 44, 244*n48*, 247*n103*
Peirce, C. S., 214
Pérez-Ramos, Antonio, 15, 16
Pinney, Thomas, 123, 126
Pittman, Charles, 45, 46
Plato, 72, 73, 75, 76, 188, 190, 198, 209, 251*n34*
Playfair, John, 12, 41, 43, 97, 98, 183
Positivism, 17, 21, 32, 35, 244*n37*. *See also* Comte, Auguste
Powell, Baden, 40, 42, 245*n73*

Realism: in geology, 9, 93, 97, 99, 100, 106, 107, 108, 120; Lewes on imagination in, 36; and Wordsworth's poetics, 54–57 *passim*; of Eliot, 122, 133, 134–35, 145, 146, 147; in Ruskin, 153, 156–72 *passim*, 179; of Sherlock Holmes stories, 232–35
Riemann, Bernard, 181–89 *passim*, 209, 263*n10*
Robertson, John, 13, 43
Romanticism: and science, 5–6, 45–53, 70, 71, 89–91, 212, 213, 250*n23*, 252*n36*; and geology, 9, 93, 100, 109–112, 122, 134–35, 145; hostility to classification in, 53, 56, 60; and *The Mill on the Floss*, 93, 100, 122, 133–34, 135, 138, 144–45; and Ruskin, 162, 163–64, 178; and Sherlock Holmes, 212, 213, 232, 234, 235
Rosenberg, John, 153, 162, 166
Royal Institution, 47, 67, 77, 78, 83, 87, 88, 169, 175, 182
Royal Society of London, 11, 12, 25, 39, 77, 83, 84, 87
Rudwick, Martin J. S., 95, 257*n7*
Ruskin, John: materialism criticized by, 9, 153, 154, 156, 162–63, 169–79 *passim*; and Tyndall, 9, 153, 154, 169–79, 262*n21*; and imagination, 9, 160–62, 164–79 *passim*; on relationship of art and science, 152–53, 154–56, 159–60; reaction by, to wave theory of light, 153–54, 172–74,

Ruskin, John (*continued*)
176–78; reaction by, to kinetic theory of matter, 153–54, 172–78; and Bacon, 163, 179; Saussure's method defended by, 166, 167–68; and methods of science, 166–179, 247*n103*
—*Modern Painters*: methodology of, 9, 152–55 *passim*, 156–166; and Baconian induction, 152–53, 156–60, 163; geology and, 152, 162, 164–71 *passim*, 176; imagination in, 153, 156, 160–66 *passim*, 167–68, 172, 176; science of aspects, 156, 162–63; materialism criticized, 156, 162–63, 172; and Romanticism, 162, 163–64; Turnerian topography described, 164–66; pathetic fallacy applied to Tyndall in, 171–72
—Other works of: *Deucalion*, 167, 168, 169, 170, 177; *Eagles's Nest*, 154, 167; *Fors Clavigera*, 169; *Love's Meinie*, 167, 169, 170; *Praeterita*, 160; *Proserpina*, 167, 169, 170; *Queen of the Air*, 153, 172, 177; "Storm-Cloud of the Nineteenth Century," 174, 175, 176
Russell, Bertrand, 184, 190, 191, 199, 209

Saussure, H. B. de, 166, 167, 168, 169
Schofield, A. T., 199
Scientific method. *See* Baconian induction; Hypotheses; Imagination; Materialism; Naive Baconianism
"Scientific use of the imagination." *See* Tyndall, John
Sebeok, Thomas, 214
Sedgwick, Adam, 94, 96, 126, 129
Sharrock, Roger, 58, 250*n18*
Shelley, Percy, 110, 111, 112, 161, 163, 251*n34*, 256*n39*
Shuttleworth, Sally, 121, 139, 141, 148, 257*n3*, 259*n32*, 260*nn39,47*
Smith, John Pye, 124, 125, 127, 128, 258*n14*
Somerville, Mary, 130
Spedding, James, 20, 27, 191–95 *passim*
Spencer, Herbert, 20, 124, 130, 139, 227
Stokes, G. G., 177–178
Sylvester, J. J., 181–89 *passim*, 197, 198, 208, 209, 210

Tallmadge, John, 118
Turner, J. M. W., 153–71 *passim*, 179
Two cultures, 5, 6, 13, 15, 43–44, 45–47, 236, 241*n2*
Tyndall, John: and "scientific use of the imagination," 8, 9, 34–35, 153, 154, 166–79 *passim*, 183, 194, 195, 200, 230, 231, 242*n12*, 246*n93*; Ruskin's criticism of, 9, 153, 154, 166, 169–79, 262*n21*; Doyle, influence on, 10, 227; importance of hypothesis for, 32–33, 168, 231–32; science and art, relation of, 37, 231, 247*n103*, 268*n26*; influence of, on Abbott, 183, 194–95, 200; and Sherlock Holmes, 230–32

*Voyage of the Beagle. See* Darwin, Charles: *Journal of Researches*

Werner, Abraham Gottlob, 97, 98, 113, 114, 116, 120
Whewell, William: on Newton, 14, 29; on hypotheses, 14, 29, 31, 34; assessment by, of Baconian induction, 14–27 *passim*, 33, 37–40, 43; idealism of, 16–17, 20, 179, 183, 188; on facts, 16–17, 20–22, 44; on Ideas, 16–17, 20–22; on imagination in science, 25, 26, 27, 29, 33, 34; and debate with Mill, 31; inductive tables, 38–39; and geology, 95, 97, 99, 100; and Ruskin, 179. Works: *History of the Inductive Sciences*, 20, 22, 25, 37, 40, 100, 179; *Philosophy of the Inductive Sciences*, 16, 22, 23, 25, 31, 37, 40, 100, 179
Wilde, Oscar, 10, 233, 234, 235, 237
Wilson, Edmund, 212
Wordsworth, William: on relation of poetry and science, 7, 8, 36, 48–55 *passim*, 60, 61, 62, 81, 91, 249*n9*, 250*n18*; on Baconian induction, 7, 47–57 *passim*, 61, 77; and De Quincey, 7, 50–53; and Davy, 8, 47, 49, 58, 77–82 *passim*, 89, 90; and Coleridge, 8, 47, 63, 68, 71, 76; and *Nature*, 45, 57; and Hamilton, 45, 61; "matter of fact," contrasted with poetry, 49–54, 71; and Darwin, 109–11, 122, 256*n42*; influence

of, on *The Mill on the Floss*, 122, 135, 138, 145; and Ruskin, 161, 164
—preface to *Lyrical Ballads*: relation of poetry and science in, 7, 8, 36, 48–49, 81, 250*n18*; responses to, by later writers, 7, 36, 49–50; naive Baconianism rejected in, 8, 49, 50; inductive qualities in, 54–56; Darwin's reading of, 112, 256*n42*
—preface to 1815 *Poems*: relation of, to preface to *Lyrical Ballads*, 8, 56–58; 60; inductive qualities of, 56–57; classification schemes of, 58–62
—Other works of: *The Excursion*, 61, 71, 256*n42*; *The Prelude*, 52, 91, 109; "Tintern Abbey," 109, 111
Wright, Willard Huntington, 212

Yeo, Richard, 15, 16, 18, 65
Young, Thomas, 25, 28

SCIENCE AND LITERATURE
A series edited by George Levine

*One Culture: Essays in Science and Literature*
Edited by George Levine

*In Pursuit of a Scientific Culture: Science, Art, and Society in the Victorian Age*
Peter Allan Dale

*Sexual Visions: Images of Gender in Science and Medicine between the Eighteenth and Twentieth Centuries*
Ludmilla Jordanova

*Writing Biology: Texts in the Social Construction of Scientific Knowledge*
Greg Myers

*Gaston Bachelard, Subversive Humanist: Texts and Readings*
Mary McAllester Jones

*Realism and Representation: Essays on the Problem of Realism in Relation to Science, Literature, and Culture*
Edited by George Levine

*Science in the New Age: The Paranormal, Its Defenders and Debunkers, and American Culture*
David J. Hess

*Fact and Feeling: Baconian Science and the Nineteenth-Century Literary Imagination*
Jonathan Smith